Fire Law

The Liabilities and Rights of the Fire Service

Thomas D. Schneid

JOHN WILEY & SONS, INC.

New York • Chichester • Weinheim • Brisbane • Singapore • Toronto

Printed in the United States of America

Library of Congress Cataloging-in-Publication Data

Schneid, Thomas D.
    Fire law: the liabilities and rights of the fire service / Thomas
  D. Schneid.
      p.   cm.
    Includes index.
    ISBN 0-471-28623-0 (hardcover)
    1. Fire fighters—Legal status, laws, etc.—United States.
  2. Fire departments--Law and legislation—United States.
  3. Liability (Law)—United States. I. Title.
  KF3976.S26   1995
  344.73'0537—dc20
  [347.304537]   99-21589

10 9 8 7 6 5 4 3 2

Contents

Foreword

Fire service ... law ... sometimes the terms seem contradictory to most of us.

In earlier times, the thought of any fire service operations being involved in a lawsuit or other litigation was almost unheard of. Fire service was almost "untouchable" in the realm of suits or other actions. That was when the "good ole boy" network existed. We can't do that anymore. The public has come to expect a degree of professionalism from all of us.

That veil of insulation no longer exists in today's fast paced society. We are responsible for our actions, and we are accountable for what we do everyday.

Litigation is everywhere, and as my old criminal law professor told me many times, "it doesn't matter whether you are right or wrong, I will defend you 'til you run out of money, and then ... "

Those are words to heed today, remembering that we are in the public spotlight each and every day, and that our customers—the public we serve—deserve the very best service that they have come to expect from the modern fire service of today.

Taking the time to do the "job" right is very important, along with documentation of what was done, or not done, as the case may be.

Tom has filled this text with some very useful information on all phases of fire service operations, government regulations, and some practical common sense that will benefit you as the provider, and the public we chose to serve.

You will have taken the first step toward making this a reality in protecting the public and yourself by reading this text on fire service law, and by following the simple and very basic tenets of the service: *to do no harm,* and to be as professional about your training, education, and activities as possible.

If in doubt, ask and learn the right procedures, and try to benefit our ultimate customer, the public. There are many volumes written on case law when things go wrong, but this text is the first step in trying to learn and prevent the loss before it occurs; i.e., to avoid closing the fire barn doors after the horse has left.

This text covers many areas of law that you may not realize affect you. OSHA, EPA, DOT, Civil and Criminal Statutes are part of your everyday potpourri of rules that are waiting out there. Remember, it is up to you as a member of the fire and emergency response service to train, educate, and document what you are doing to protect your members, employees, and the customers you serve.

Education is the key to survival in the future. Your future depends on the lessons that you learn and heed from this textbook.

Safety is the responsibility of everyone involved. Education is also the responsibility of everyone in the fire and emergency services.

Remember, any job worth doing is worth doing right. Learn, train, and prepare for your important job in the most rewarding field of fire and emergency services.

Michael J. Fagel

Preface

This text was prepared specifically for fire service personnel, using the most up-to-date information available. This text is designed to provide a general overview of selected areas of the law applicable to the fire service and to provide appropriate case analysis to assist the reader in interpreting the law. Additionally, where appropriate, guidelines or interpretations are provided to assist the reader in achieving compliance with the applicable law.

While every effort has been made to ensure the complete and total accuracy of law, the law is a constantly evolving subject. The author provides no warranty, either expressed or implied, as to the law in each of the categories. Although suggestions are offered, the author does not intend this text to provide specific counsel with regard to all possible circumstances. Competent local legal counsel should be acquired to assist in specific circumstances and situations.

This text is offered to fire service personnel in the hope that a greater understanding and appreciation of the law as it applies to the fire service can be achieved. Given the growing complexity of the law, fire service personnel should know their rights, and those charged with the management of fire service personnel must ensure that these rights are protected, and that compliance is achieved and maintained to protect your fire service organization. If knowledge is power, it is the author's hope that this text will play a small role in empowering fire service personnel within the legal arena.

TO THE READER

This text is designed to cover four major areas of the law: (1) overview and foundational elements, (2) potential civil liabilities, (3) potential criminal liabilities, and (4) potential administrative liabilities. Within each of these major areas, specific potential liabilities are explored through a generalized view of the legal requirements and case studies.

The reader should be aware that not all areas of potential liability for fire fighters are covered in this text. The author attempted to identify the areas that have the greatest frequency of occurrence or carry the greatest potential severity in terms of punishment for failure to comply with the law. Each of the major areas is presented in general, and then specific areas of importance to fire fighters are addressed with specificity. Additionally, each major area and chapter contains one or more cases to assist the reader in understanding the basis for the law and the judicial enforcement by the courts.

READING A COURT DECISION

The following suggestions may be helpful with regard to the court decisions or cases provided in this text:

- Determine what type of case you are reading, i.e. civil, criminal, and so on.
- Look at the title of the case and which court is making the decision, i.e., U.S. Supreme Court, Madison County District Court, and so forth.
- Note the date the decision was handed down by the court.
- Read the brief summary or headnote, usually provided at the beginning of each decision.
- Note whether the decision was unanimous or handed down by a divided court. Note whether a dissent was written by the judges who did not agree with the majority decision.
- Read each decision fully at least once to get a "flavor" for the case. Read the decision again slowly and with specificity.

- Look up legal terms you do not fully understand in a legal dictionary.
- Brief your case in order that you fully understand the issues, facts, and decision in the case.

Analyzing Case Law and Briefing a Case

Case law is the accumulation of court decisions which, in essence, shape and develop new law. Throughout this text, you will find numerous cases that will identify and exemplify the particular point of law and provide the court's decision-making process as to how the judges derived their answer in the particular case.

As a basic rule of thumb, when analyzing and/or briefing a case, you should read through the case in detail once. On the second reading, you should identify the basic issues, facts, and holding of the decision, as well as any dissenting opinion. On the third reading of the case, you should take notes with regard to the actual brief of the case to which you will be referring at some later point in time. It is essential that you take good notes and brief your case extremely well in order that you may refer to the brief and refresh your memory later in your studies.

Finding the Case

Finding the case in the library is often one of the most difficult parts of your total analysis. As can be seen from the various cases provided in this text, other cases are referenced, and numerous "cites" are provided throughout this text. All reported decisions of cases in the United States judicial system are listed according to the publication in which the case appears (called a reporter), the volume in which the case is located, and the initial page number. Take, for example, 36 S.E. 2d 924. If you were searching for this case, you would go to the South Eastern Reporter, second edition, look for volume number 36, and find the case at page 924. In the federal judicial system, the publications tend to be in accordance with the region of the country, and in most state judicial systems, the location will be in a state reporter. It should be noted that not all cases are reported and published. Generally, trial court decisions are not published, because these decisions do not serve as mandatory precedent for other courts to follow. Usually, only decisions of federal and state appellate courts are published.

Statutes, regulations, and standards tend to be within other publications such as the *Code of Federal Regulations*. This system is set up using the same publication, volume, and page number as with the judicial court decisions. However, the series numbers may reflect the particular regulation or standard. For example, 29 C.F.R. 1910.120 is the Occupational Safety and Health Administration's Standard with regard to hazardous waste operation and emergency response. If you were searching for this particular standard, you would go to the *Code of Federal Regulations,* which is signified by the C.F.R. designation, volume 29, and section 1910.120.

In addition to the normal procedure in locating a case, some law offices and law libraries also provide an electronic database through which to locate cases. The two major databases are WESTLAW and LEXIS. Each of these databases normally provides training and publications to assist you in locating a particular case. As a general rule, these databases provide a basic menu of the various areas where the law is located and numerous sub-databases that guide your search. For example, if you are searching for a federal decision, you may enter the federal database and then narrow your search to the particular area of law that you are seeking. If you should possess the case name, the case name normally can be used to pull up the case. If you are searching for a particular issue of the law, these databases will provide the case cites for your review and evaluation. It is highly recommended that you acquire the particular training or assistance from the librarian at the law library before conducting any search on WESTLAW or LEXIS.

It is highly recommended that you become familiar wit.h the particular library that you will be using during this course. Please take a few minutes out of your busy schedule and walk through the library to locate the different publications that are available, and note the location of each of these publications. Please thumb through a few of the publications and test your skill at locating particular cases to be able to find such cases efficiently in the future.

Briefing the Case

The basic purpose of briefing a case is to help you understand the particular legal issues of the case and their significance. There are various methods of briefing a case, and the following format is only offered as an example of one of these methods. Your instructor may suggest an additional format, or you may devise your own system to help analyze these cases. No matter what method you adopt, read the case thoroughly at least once to get a general idea of what the case is about before beginning to take notes for your brief.

The basic recommended framework for briefing a case includes the following:

1. issues
2. facts
3. holding (decision)
4. dissent
5. your opinion
6. underlying policy reason issues

Identify the basic issue or issues that are in question before the court. To find the basic issue or issues involved, you have to identify the rule of law that governs the dispute and ask how it applies to the particular facts of the case. In most circumstances, you will be writing the issue for your case brief as a question that combines the rule of law with the material facts of the case. For example, does the arson statute in the state of Kentucky apply to a minor child?

Facts

The facts of the particular case describe the events between the conflicting parties that led to this particular litigation and tell how the case came before the court that is now deciding it. Often included in the facts are the relevance to the issue the court must decide and the basic reasons for its decision. When you first read through the case, you will not know which facts are relevant until you know what the issue or issues are in the particular case. Thus, it is vitally important that you read through the case at least one time before beginning to summarize the facts.

In addition to the specific facts of the situation, it is important to see what court decisions have come prior to the case which you are currently reviewing. Often, the published decisions are appellate decisions, and thus a district court or circuit court has decided the matter previously, and the matter is now on appeal. If the particular facts of the situation in an appellate case are not provided in detail, you may want to research and review the district or circuit court decisions to acquire the particular facts in your case.

In this section, you should also include the relevant background for this case. You should identify who the plaintiff and the defendant are, the basis of the plaintiff's suit, and the relief the plaintiff is seeking. You may also want to include the procedural history of the case such as Motions to Dismiss and other motions that are relevant to the case. In an appeals case, the decisions of the lower courts, grounds for those decisions, and the parties who appealed should also be noted.

Within this facts section, you should be as brief as possible. However, all pertinent points should be noted. Although this is a judgement call, most statements of facts in a brief should not be more than two or three paragraphs in length. Given the fact that you will have read the case at least three times while briefing the case, the facts provided in your brief of the case should be the major points used to refresh your memory.

Holding

The holding is the court's decision on the question that was actually presented before it by the parties. The court may make a number of legal statements but, if it does not relate to the question actually before it, this information is considered dicta. The holding can normally be identified by the statement "the court has decided . . ." or "the majority decision is . . . ." A holding, in essence, provides the answer to the question you were asking in your issue statement. If there is more than one issue involved in the case, there may be more than one holding in any given case.

Dissent

Often, with U.S. Supreme Court cases and appellate cases, the majority decision is the decision of the court. However, the minority is also provided an opportunity to give their reasoning as to their dissent in the decision-making process. Although the dissenting opinion is not law and has no bearing on the case, the dissent provides another point of view on the particular issue of the case and also may be referred to in some later case.

Opinion

After you have reviewed the case at least three times and have analyzed the court decision and briefed your case, you should have a good idea whether you agree or disagree with the court's opinion. In this section, you should provide your personal opinion as to whether you agree or disagree with the court and why.

Policy

In many cases, there is an underlying social policy or particular social goal that the court wishes to further. When a court explicitly refers to those policies in a particular case, this information should be included in your brief because it will provide to you a better understanding of the court's decision. For example, in the historic case of *Brown v. Board of Education,* the decision of the court was formed through an underlying social policy to eliminate discrimination in our school system. This underlying social policy is often very important in appellate and Supreme Court decisions. Following is an example brief for your review and evaluation. It is highly recommended that you "test" your skills by briefing several of the cases within this text or other cases before your initial briefing of a case for a class.

In addition to the above stated information, below are several other helpful hints that may assist you in briefing the case:

1. Try to confine your brief to a maximum of one page. If your brief is over two pages, you have probably provided too much information. Remember, a brief is to refresh your memory at the time you need to recall this information for class or other purposes.

2. The case that is printed in the textbook may have been edited and shortened by the author of the textbook. Normally, a full court opinion may run 20 or more pages; thus, the case has been shortened for the purposes of the text. If you find that you are having difficulty understanding the case because so much information has been deleted, you may want to go to the library and read the full text opinion.

3. During your first couple of attempts at briefing a case, it is often difficult to extract the important elements and issues. Please keep in mind that not all judges are expert writers, so the opinions may often be confusing or difficult to understand. Additionally, you should realize that not all courts follow the same format in writing opinions, so you may find some decisions more difficult to understand than others. Therefore, you may find that judges sometimes go off on a tangent and discuss other rules and points of law that are not essential to the determination of the particular case. It is your job to filter through this information to identify the particular issues and laws that are applicable to the case.

4. You may often run across Latin or "legal" terms with which you are not familiar. Since you will need to have a clear understanding of the terminology used in the particular case, you will have to look up the terms in a legal dictionary. It is often a good idea to have a *Black's Law Dictionary, Ballantine Law Dictionary, Gilmer's Law Dictionary,* or other law dictionary

available while you are reading and briefing the case. Standard dictionaries often do not include these terms, or the explanation provided may be incomplete. Use of a law dictionary is essential when reading these cases.

5. When reading the cases provided in this text, you may want to look at the particular chapter and section headings of the textbook in which the case appears. If you are having difficulty identifying the particular issue of the case, the issue is normally related to the topic discussed in the chapter or section heading. The cases in this text have been inserted to illustrate the subject matter being discussed in each of the chapters.

6. Remember, the issue or issues in the particular case should always be stated in the form of a question. You should never begin your issue with the words "whether or not" because this will not form an interrogatory question. Also, the terminology "should plaintiff win" or "is there a contract" are not correct forms of stating the particular issue.

7. In determining the particular rule of law, ask yourself, "If I had to tell someone who knew nothing about this case what this case is about or what it stands for in one sentence, what would I say?" Often, the rule of law can be determined by taking the issue and putting it in the form of a declarative statement and adding a few words. For example, in the case of *Miranda v. Arizona*, 384 U.S. 436 (1966), the issue and rule may be as follows:

Issue

When a person is taken into police custody or otherwise deprived of his freedom of action in a significant way, must his constitutional rights to remain silent and to have an attorney present be explained to him prior to questioning?

Rule

When a person is taken into police custody, the following warnings must be given prior to questioning:

a. He has the right to remain silent.

b. Any statement he makes may be used against him as evidence.

c. He has the right to have an attorney present.

d. If he cannot afford an attorney, one will be appointed for him.

8. Finally, do not use other people's briefs. Without having read a particular case and analyzed the court decision yourself, use of another individual's case brief is essentially worthless. A brief is simply the codification of your thoughts and work to which you will refer in the future to refresh your memory.

Example Case Brief

Case Name: *Marshall v. Barlow's, Inc.*, 436 U.S. 307 (1978)

Issue: Is Section 8(a) of the Occupational Safety and Health Act unconstitutional in that it violates the Fourth Amendment?

Facts: Appellee (Barlow's, Inc.) initially brought this action to obtain injunctive relief against a warrantless inspection of its business premises by Appellant (Secretary of Labor Marshall). The inspection was permitted under Section 8(a) of the Occupational Safety and Health Act of 1970 which authorizes agents of the Secretary of Labor to search the work area of any employment facility within OSHA's jurisdiction for safety hazards and OSHA violations without obtaining a search warrant or other process. A three judge Idaho District Court ruled in favor of Barlow's and concluded that the Fourth Amendment required a warrant for this type of search and that the statutory authorization for warrantless inspections was unconstitutional. This appeal resulted.

Holding: Yes, Section 8(a) of the Occupational Safety and Health Act of 1970 was unconstitutional in that it violated the Fourth Amendment. The U.S. Supreme Court affirmed the decision of the Idaho District Court and granted Barlow's an injunction enjoining the enforcement of the act to that extent.

        The court states that the rule against warrantless searches applies to commercial premises as well as private homes. Although an exception to this rule is applied to certain "carefully defined classes of cases" including closely regulated businesses such as the firearm and liquor industries, this exception does not automatically apply to all businesses engaged in interstate commerce, as the Secretary alleges.

Opinion: I agree with the court in this case. I feel that requiring search warrants insures that the search is a reasonable one and that the particular business being inspected is not merely being singled out (for one reason or another). I agree with the court in that requiring search warrants will not make inspections less effective nor will it prevent necessary inspections, but rather will serve to insure fairness in inspections.

Policy: Although no specific public policy was mentioned in the case, the implied policy was that of pro-business, anti-regulation.

West's National Reporter System

1. *Atlantic Reporter*—includes cases from the following states: Maine, Vermont, New Hampshire, Connecticut, Rhode Island, New Jersey, Maryland, Delaware, and Pennsylvania.
Cited as A. or A.2d

2. *North Eastern Reporter*—includes cases from the following states: Illinois, Indiana, Ohio, New York and Massachusetts.
Cited as N.E. or N.E.2d

3. *South Eastern Reporter*—includes cases from the following states: Georgia, South Carolina, North Carolina, Virginia and West Virginia.
Cited as S.E. or S.E.2d

4. *Southern Reporter*—includes cases from the following states: Louisiana, Mississippi, Alabama and Florida.
Cited as So. or So.2d

5. *North Western Reporter*—includes cases from the following states: North Dakota, South Dakota, Nebraska, Iowa, Minnesota, Wisconsin, and Michigan.
Cited as N.W. or N.W.2d

6. *South Western Reporter*—includes cases from the following states: Texas, Missouri, Arkansas, Tennessee and Kentucky.
Cited as S.W. or S.W.2d

7. *Pacific Reporter*—includes cases from the following states: Washington, Oregon, Montana, Idaho, Wyoming, California, Nevada, Utah, Arizona, New Mexico, Colorado, Oklahoma and Kansas.
Cited as P. or P.2d

Other Reporters

United States Reports (U.S.) or *Supreme Court Reporter* (S.Ct.)
   contain U.S. Supreme Court Cases
Federal Reporter (F. or F.2d.)
Federal Supplement (F. Supp.)
Federal Cases (F. Cas.)

Citation Forms for Rules, Regulations, and Standards

1. Federal Rules and Regulations (except Treasury):
   Code of Federal Regulations—C.F.R.
   Federal Register—Fed. Reg.

2. Formal Advisory Opinions:
   Opinions of the Attorney General—Op. Att'y Gen.
   United States Statutes:
   United States Code—U.S.C.
   United States Code Annotated—U.S.C.A. (West 19xx) (a
      West Publication)
   United States Code Service—U.S.C.S. (Law. Co-op.
      19xx) (a Lawyers Cooperative Publication)

3. Other United States Official Administrative Publications:
   *Administrative Decisions under Immigration and Nation-
      ality Laws*—I.& N. Dec.
   Agricultural Decisions—Agric. Dec.
   Civil Aeronautics Board Reports—C.A.B.
   Copyright Decisions—Copy. Dec.
   Cumulative Bulletin—C.B.
   Customs Bulletin and Decisions—Cust. B. & Dec.
   *Decisions and Orders of the National Labor Relations
      Board*—N.L.R.B.
   Decisions of the Commissioner of Patents—Dec.
      Comm'r Pat.

Decisions of the Comptroller General—Comp. Gen

Decisions of the Employees' Compensation Appeals Board—Empl. Comp. App. Bd.

Decisions of the Department of the Interior—Interior Dec.

Decisions of the Federal Labor Relations Authority—F.L.R.A.

Decisions of the United States Merit Systems Protection Board—M.S.P.B.

Federal Communications Commission Board Reports—F.C.C. or F.C.C. 2d

Federal Energy Regulatory Commission—F.E.R.C.

Federal Maritime Commission Reports—F.M.C.

Federal Mine Safety and Health Review Commission Decisions—F.M.S.H.R.C.

Federal Reserve Bulletin—Fed. Res. Bull.

Federal Service Impasses Panel Releases—Fed. Serv. Imp. Pan. Rels.

Federal Trade Commission Decisions—F.T.C.

Interstate Commerce Commission Reports—I.C.C.

Interstate Commerce Commission Reports, Motor Carrier Cases—M.C.C.

National Railroad Adjustment Board—N.R.A.B.

National Transportation Safety Board Decisions—N.T.S.B.

Nuclear Regulatory Commission Issuances—N.R.C.

Official Gazette of the United States Patent Office—Off. Gaz. Pat. Office

Opinions of Office of Legal Counsel of the Department of Justice—Op. Off. Legal Counsel

Securities and Exchange Commission Decisions and Reports—S.E.C.

Social Security Rulings Cumulative Edition—S.S.R. (Cum Ed. 19xx)

Acknowledgments

My undying thanks to Ms. Nadia Rawlins, Mr. Howard Hallinan, Mr. George Houghton, and Ms. Krista Reynolds for their assistance in the preparation of this text. Without their help, this task would have been overwhelming.

Special thanks to my wife, Jani, and my daughters, Shelby and Madison, for putting up with me during the writing of this text.

1

Overview of the Law

Law is nothing unless close behind it stands a warm. living public opinion.

Wendell Phillips

What are laws? Where do laws come from? Who makes the laws? Does Congress make the laws, or is it the President? In the United States, we use the terminology "law" to encompass a wide variety of rules and regulations which govern our conduct on a daily basis. A "law" can mean the rules established by Congress, by our state legislatures, by the courts, or it can be construed as common practices which apply to everyone within our community. Hence, if we did not have rules and regulations to govern the conduct of our citizenry, the result would be chaos. Laws govern our conduct in order that individual rights are preserved and that commerce can run efficiently. Defined in *U.S. Fidelity and Guarantee Company v. Guenther*, "law is used in a generic sense, as meaning the rules of action or conduct duly prescribed by the controlling authority, and having the binding force of law." In the United States, there is a wide variety of laws ranging by our conduct. Laws may be civil in nature (i.e., money damages), criminal in nature (i.e., jail time), administrative laws (i.e., Occupational Safety and Health Administration Regulations), general laws, and numerous other types of laws. We should additionally understand that the laws in the United States are constantly evolving and changing in order to encompass the new technologies and the new rules of conduct as our society grows.

Laws generally consist of four major parts:

1. the declaratory aspect of the law where the rights to be observed or the wrongs to be avoided are clearly defined and published
2. direct pre-phase where the individuals involved are instructed to observe the particular rights and abstain from the commission of the wrongs involved
3. the remedial phase points out the method to recover an individual's private rights or to redress private wrongs
4. the sanction phase in which the punishments for violation of the law are outlined for all to review

In the United States, individuals must be provided knowledge of the law before it can be enforced. Individuals are required to know what the law is, and it has been said numerous times, "ignorance of the law is no defense."

SOURCES OF THE LAW

If you walk into any law library, you will notice the vast volumes of laws and decisions that are currently enforced in the United States. These are normally made by the Legislative Branch of the Federal Government or the State Government, but they can also be made by the Executive Branch and interpretations of the law which, in actuality, can be law in and of themselves. The basis for most laws in the United States is the U.S. Constitution. The U.S. Constitution contains the fundamental law or principles upon which the United States is based. Congress and state legislatures normally pass statutory laws with approval of the Executive Branch. Laws can also be passed by Federal Agencies and State Administrative Agencies. Ordinances are usually developed by cities or county municipalities. Other "laws" can be developed within the boundaries of the U.S. Constitution. It should be noted that there is no one single source which defines all of the laws in the U.S. The law is basically a compilation of documents and laws which have been interpreted by the courts to be Constitutional. This constitutes the overall "law" in the United States.

CONSTITUTIONAL LAW

The basic foundation of all laws in the United States is the U.S. Constitution. This Constitution is the "law of the land" as developed by the founding fathers of the United States of America. The Constitution is the highest law, above all statutory and decisional law rendered by any state, federal, or local entity. The rights guaranteed by the Constitution are spoken of as a constitutional right, such as the freedom of speech and the freedom of press as outlined in the First Amendment of the U.S. Constitution. The rights established under the U.S. Constitution can not be abolished by the federal or state government, or by any municipality or any political subdivision. Constitutional rights are fundamental rights in the United States but are not considered absolute.

Most state constitutions are modeled after the U.S. Constitution. Any state constitution must adhere to the principles outlined in the U.S. Constitution and any state or local law or regulation may not oppose requirements set forth under the U.S. Constitution. This constitution has established that certain powers belong to the state government and that others belong to the federal government. These powers can not be revoked. However, in certain areas the federal and state laws may overlap.

COMMON LAW

The law dictionary defines common law, "distinguished from law created by the enactment of legislatures, the common law comprises the body of those principles and rules of action, related to the government and security of persons and property, which derive their authority solely from usage and custom of immemorial antiquity, or from the judgment or decree from the court recognizing, affirming, or enforcing the usage and customs; and in this sense, particularly the ancient unwritten laws of England. The "common law" is the statutory and case law background brought from England to the American colonies before the American Revolution. Such law consists of those principles, usage and rules of persons and property which do not rest for their authority upon any expressed or positive declaration of the will of the legislature.

In a broad sense, common law may designate all that part of the positive law, common juristic theory, an ancient custom of any state or nation which of general or universal application, thus marking off special or local rules or customs. In the United States, what we refer to as common law is often the ordinary customary usage brought down for years from our ancestors. In a vast majority of states, common law has been adopted by statutes. However, common law is generally determined only by the review of all court decisions on a particular subject in question. If a court repeatedly affirms a specific principle, such as at-will employment, the rule takes on the status of law. Decisions of the court over a period of time are judicial precedents set by the court and in essence become law."

STATUTORY LAW

Laws are passed by the U.S. Congress or by state legislature and, if they are not vetoed by the executive branch or found unconstitutional by the judicial system, are statutory law. Statutory law includes all laws and other enactments passed by any legislative body and are normally codified in codes or statutes. Statutory law is subject to revision with deletion by the legislative body at any time. Statutory law may address a wide variety of subject areas within the jurisdiction of a legislative body. Federal law is normally codified in the United States Code (U.S.C.) and most states have a particular version of the statutes (such as the Kentucky Revised Statutes—K.R.S.).

The widely diverse areas covered in most statutory law is divided into particular areas. For example, most states have the Health and Safety Code in one particular area with statutes and provide supplements for any changes that are made during a legislative session. These laws are of particular importance to a fire service organization. The requirements for the formation of a fire district or a fire corporation are included in the state statutes. Laws regarding vehicle code registration, labor code, safety and health regulations, and related laws are normally codified as statutory law.

Statutory law can also be at the local level. Communities or counties frequently pass ordinances which regulates such areas as construction, operation, and building codes. Regulations are usually compiled and published in one document such as the Communities Municipal Code.

ADMINISTRATIVE LAW

An area of particular importance for fire service organizations is administrative law. Given the substantial number of governmental agencies which have been created by various laws enacted by Congress or state legislatures, the administrative agency or commission provided the responsibility in enforcing a particular statute which normally creates administrative rules through which to clarify, implement, or enforce the particular law. In the bounds of the statutory power granted to them, these agencies are given wide discretion with the regards to development of specific requirements and implementations of procedures of achieving compliance with the particular statutory law. Administrative agencies, such as OSHA, EPA, FAA, and even state fire marshall offices, are considered to have quasi legislative and quasi judicial powers. These agencies are normally permitted to promulgate technique rules, regulations, procedures, so they normally have broad enforcement powers. Governmental agencies use their administrative rules and these rules are enforced in a variety of manners.

CASE LAW

Case law and common law are sometimes used interchangeably by the general public however there is a distinct difference. Case law, however, refers to the decisions made by the courts in various cases ranging from the U.S. Supreme Court through the local district court. As cases are appealed through the judicial systems, whether federal or state, the decision of the lower court will be heard by a different set of judges or by the same court based on decisions of different reasoning or different public need. The decisional law is always subject to change on appeal. Although the uniform law on a number of subjects has been adopted by various states, there is no obligation the states to adopt the uniform law. This can result in the statutory law in one state being exactly opposite in another, despite identical facts and circumstances. Sometimes, appellate courts in different districts of the same state do not even agree. When this happens, the law in the state remains uncertain until the matter is appealed to a higher court within the state or ultimately to the U.S. Supreme Court.

Whether the case is brought in federal or state court it can have a major difference. Because state courts run as independent sovereign judicial bodies, the state court may have different procedure requirements, different rules of evidence, and often state courts find different opinions on the same legal question. The courts or "the forum" is vitally important. Prosecutors often seek to try the case in a court that is more likely to hand down a decision favorable to his/her client. Please note that there are specific rules regarding the category of conflicts of law in deciding which court should hear a particular case. (See Conflicts in Law.)

LAWS DIRECTLY AFFECTING THE FIRE SERVICE

In the past, fire service organizations were often immune from prosecution or civil actions under specific laws provided usually by state governments under the common law theory that the "king can do no wrong" known as sovereign immunity. With the changing of the times and the laws, fire service organizations are now viewed by many as being another "deep pocket" for civil actions and fire fighters are subject to the same criminal laws as the average citizen. Fire fighters should be aware that they are not exempt from any laws and are required to comply with the laws as any other citizen in the United States. Fire fighters can be held liable for any criminal actions that they may be involved in while on or off the job. Fire service organizations can be held liable for not only the actions that they do but also for the omissions when a duty has been created. In short, fire service organizations must know what laws apply to their particular organization and their fire fighters must comply with these laws. Failure to comply can result in the same types of sanctions, whether criminal or civil, as with any other individual or organization in the United States.

In the criminal arena, fire fighters should be especially sensitive in the area of the criminal violations for arson. As exemplified in the movie *Backdraft*, fire fighters are responsible for compliance with the law whether they are on the job or off the job. As discussed later in the criminal sections of this text, fire fighters are held liable for such criminal violations in the course of their employment. In the civil action arena, fire service organizations are quickly becoming one of the "deep pockets" for potential plaintiffs. Fire service organizations are now being scrutinized as to procedures going to and at the fire, and during salvage and clean-up operations and in all aspects of the activities at the fire scene. Discussed later in this text, such situations as the 911 communications system, the tactics, and omissions by fire fighters at the fire scene are becoming the basis for negligence actions against the fire service organization.

It was a well established principle of law that fire fighting and fire service organizations are considered a a government undertaking. This means the fire fighting activity was performed effectively only by an agency of the government. The intended benefits conferred on a fire department was for the community as a whole (i.e., to prevent a fire or to extinguish of a fire) and not for the protection of individual interests. Therefore, the duty owed by a fire department was considered limited in the past and restricted to protecting the whole community. In performing the total community or governmental function, fire departments were considered immune from liability from loses of an individual, whether it occurred by the acts or omissions or commissions of the fire service organization. Under the concept that "the king can do no wrong," fire service organizations in the past were normally immune from civil actions by a private individual for harms which occurred in activities at the fire scene or in related activities. The concept of immunity for fire service organizations is rapidly deteriorating in the law, creating a greater risk of liability to fire service organizations and the fire officers and fire fighters within that organization.

2

Organization of the Judiciary System

The aim of law is the maximum gratification of the nervous system of man.

Justice Learned Hand

The American judicial system is composed of numerous levels of courts encompassed within two broad categories, namely the federal judicial system and the individual state or territory's judicial system. Jurisdiction and the specific rules governing a particular court dictate which court has the authority to render an enforceable decision over the particular matter in dispute.

Jurisdiction by the courts over any given matter can be determined in two basic manners: (1) jurisdiction over the person or (2) jurisdiction over the object. In essence, for a court to be able to hear and make an enforceable decision over a particular matter, the case must involve a resident of the state or the situation (i.e., the automobile accident must have occurred within the state). Some courts, such as tax courts and traffic courts, have special jurisdiction over special matters.

A particular court's jurisdiction over an issue can also be limited by the type of case, the dollar amount involved, and other factors. In most states, specific jurisdictional requirements are established by the courts or legislature. These jurisdictional requirements can usually be found in the state's court rules or in the state statutes.

In the federal court system, the offense must be within the jurisdiction of the federal courts or diversity must exist in order for the federal court to be able to hear the case. For example, if an individual committed arson involving a federal building, the matter is a federal offense and the case would be heard in the federal court system. Conversely, if a West Virginia resident causes damage to a building owned by a West Virginia resident and the building was located in West Virginia, the West Virginia state court system would have jurisdiction. However, if the building was located in another state, or the parties were not residents of West Virginia, diversity may exist permitting the federal court system to acquire jurisdiction. It should be noted that federal courts can adjudge cases involving questions of state law and state courts can consider cases involving federal law.

Jurisdiction is also classified as original or appellate. The court which originally hears the case is considered to have original jurisdiction. If the decision of the original court is appealed to a higher court, the higher court must have appellate jurisdiction in order to hear the appeal. In courts having original jurisdiction, usually known as trial courts, district courts, or circuit courts, the "Perry Mason" or "L.A. Law" type of hearings are held. In the court of original jurisdiction, evidence is presented, witnesses are called, arguments are made, and the court determines the matter of a factual basis. In courts having appellate jurisdiction, the determination is usually made with limited, if any, oral arguments, and based on the record provided by the court of original jurisdiction. The appellate court reviews the judgment of the court of original jurisdiction to ensure that all appropriate rules and procedures were followed and that there was appropriate factual evidence to support the decision of the jury or judge in the case. If the rules were followed and the facts substantiated the decision, most appellate courts are required to sustain the decision of the court with original jurisdiction.

In determining which court is appropriate for hearing a particular case at the original jurisdictional level, the first determination is who are the parties involved and where was the "situs" or location of the incident or offense in dispute. The second determination should be which court system, federal or state, has original jurisdiction over the matter. The third level of determination should be which court having jurisdiction over the matter because of the amount of money involved or other specific requirements. The determination of the appropriate court is vital in insuring that the issue in dispute is properly adjudicated.

FEDERAL JUDICIARY SYSTEM

Article III, Section 1 of the U.S. Constitution provides:

The judicial power of the United States, shall be vested in one Supreme Court, and in such inferior courts as the Congress may from time to time ordain and establish.

The federal judicial system is composed of the U.S. Supreme Court, twelve circuit courts of appeal, district courts, and various special courts such as territorial courts and U.S. Tax Court. Additionally, there is a specialized circuit court system known as the

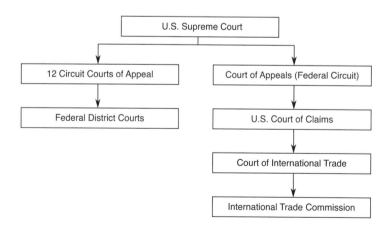

U.S. Court of Appeals for the Federal Circuit, which has exclusive jurisdiction to hear all appeals from the U.S. Court of Claims, the Court of International Trade and International Trade Commission.

SUPREME COURT OF THE UNITED STATES

The United States Constitution designates the Supreme Court to be the highest court in the United States. The Supreme Court has original jurisdiction over cases where one state is suing another, where courts of appeal have differences of opinions with regard to a federal issue, or in hearing appeals from the Circuit Courts of Appeal. The jurisdiction of the Supreme Court is limited to cases involving international law, federal law, and federally protected rights. If none of the issues involve federal law or international law, the U.S. Supreme Court would therefore lack authority to consider the case and must defer to the judgment of the state supreme courts. The Constitution does not expressly grant the Supreme Court the authority to declare a law that Congress or a state legislature has passed to be invalid. The famous case of *Marbury v. Madison,* found that since the Constitution is the supreme law in the United States, a court may hold invalid any attempt to circumvent the provisions of the U.S. Constitution. Just as John Marshall stated in *Marbury v. Madison,* the U.S. Supreme Court did not have this authority and that it would nullify the values of the written Constitution and permit it to be altered by any statute which the current Congress or state legislature would happen to pass.

The U.S. Supreme Court is the highest appellate court in the United States. Some cases are required to be heard by the U.S. Supreme Court by right. A vast majority of cases are brought up on *certiorari*. Writs of *Certiorari* to the U.S. Supreme Courts are basically requests for discretionary review of a lower court's ruling. Since the ruling of the U.S. Supreme Court is the final rule with most decisions in the United States, the only method in which a

U.S. Supreme Court decision can be modified or overturned is with an act of Congress (i.e., The Civil Rights Act of 1990).

CIRCUIT COURTS OF APPEAL

This judicial system is divided into 12 federal judicial circuits within the United States. Courts of Appeal consider all appeals from the District Courts within their jurisdiction or district unless the case is one in which the appeal could go directly to the U.S. Supreme Court. Each circuit encompasses between three and nine states and the U.S. Court of Appeals hears all appellate decisions from district courts within that region. The U.S. Courts of Appeal establish legal precedents which all United States District Courts within that particular circuit must follow. Law on a particular subject may vary among and between the circuits with resolution of any conflict coming only through appeal to and decision by the U.S. Supreme Court.

FEDERAL DISTRICT COURTS

District courts are the trial level courts of the federal judicial system. These are the courts in which witnesses will be heard, testimony will be taken, and an initial decision rendered. As noted in the diagram on page 11, the Federal District Court is the initial or entry level in most federal cases and all levels of courts above the Federal District Court are considered appellate. District courts hear both criminal and civil actions.

SPECIAL COURTS

Federal District Courts normally have their own rules and procedures to be followed in five special courts. Within the federal judiciary system, special courts have been developed to address particular issues. At the circuit court level, there is a U.S. Court of Appeals for the Federal Circuit which is based in Washington, D.C. and provided exclusive jurisdiction for appellate decisions from the U.S. Court of Claims, Court of International Trade, and International Trade Commission. In the area of tax, a special United States Tax Court has been developed for any appeals from decisions made by the Commissioner for the Internal Revenue Commission. The United States Tax Court does not handle claims for overpayment of taxes. These matters are normally handled by the U.S. Court of Claims or U.S. Federal District Court.

STATE COURT SYSTEMS

State Court systems are usually structured on the same basic arrangement as the federal court system. However, it should be noted that the names used for each of these courts may vary from state to state. (For example, the New York Supreme Court is the trial level court.) State court systems are typically composed of a state appellate court, district courts of appeal or circuit courts of appeal, courts which may be called superior or supreme courts, and smaller specialized courts such as municipal courts, traffic courts, or justice courts.

The state supreme court is normally the highest appellate court within the state. Particularly, the state supreme court is made up of a chief justice and anywhere from 4 to 8 associate justices. The state supreme court normally hears all appeal cases from the lower state courts or cases involving interpretation of state law from a federal court. Cases involving the death penalty, *habeas corpus*, or other criminal related matters are heard by the state supreme court.

States have several courts of appeals which are responsible for a designated area within the particular state. These courts normally hear all of the appeal cases within that state. The court of appeals normally considers appeals from the trial court judgments and orders unless the state supreme court assumes jurisdiction over the matter. The entry level into most state court systems is the trial court or district court level. Courts can be called various names such as superior court, supreme court or district court. District courts are the initial determining body in most civil and criminal matters. State district courts have jurisdiction to consider other cases involving such things as individual taxes, probate, domestic relations and juvenile problems. District courts or trial courts are also the courts having jurisdiction over most criminal matters involving felonies and particular misdemeanors within the state.

MUNICIPAL COURTS AND OTHER COURTS

States with large populations have established municipal courts, traffic courts, and other minor courts to handle particular matters. In some jurisdictions, municipal courts or small claims courts handle civil cases which do not involve a large amount of money. In other jurisdictions, municipal courts have been established to hear cases involving misdemeanors except those involving juveniles. In these states, special courts have been established to hear juvenile offenses.

In some states, courts have been established with special limited jurisdiction to hear particular matters. Such courts include city courts, city justice courts, and township courts. These courts are normally established to assist another court when there is a particular specialized problem, such as building and fire code violations, where specialized training is needed by the judge or court official.

In summation, the court systems in the United States are designed to act as referees between two or more adversarial parties. Specialized roles and procedures have been established for both criminal and civil proceedings. The U.S. Constitution and other laws within our judicial court system guarantee citizens a means by which to redress any injustice or wrong that has been done to them. Individuals must assert their rights in a civil action in order to be able to achieve justice—and the state, acting on behalf of the citizenry, must initiate criminal actions.

3

Civil and Criminal Law

The law must be stable and yet it must not stand still.

Roscoe Pound

DISTINCTION BETWEEN CIVIL AND CRIMINAL LAWS

In the Unites States judicial system, the same courts may hear a case whether the issue involves a civil or criminal matter. The differences, however, between a civil matter and a criminal matter are significant. According to *Black's Law Dictionary*, civil law is body of law which every particular nation, commonwealth, or city has established particular to itself; more properly called "municipal" law, to distinguish it from "law of nature" and from international law. Laws concerning civil or private rights or remedies, as contrasted with criminal law.[1] Criminal Law is defined as "The substantive criminal law is that law which for the purpose of preventing harm to society, (a) declares what conduct is criminal, (b) prescribes that punishment to be imposed for such conduct. It includes the definition of specific offenses and general principals of liability. Substantive criminal laws are commonly codified into criminal or penal codes . . ."[2] In essence, civil law can be thought of as the private individuals pursuit of redress, usually money damages, while criminal law deals with harms against society, and the redress can be the removal of personal freedoms, i.e., jail.

---

[1]*Blacks Law Dictionary*, West Publishing Company (1983).
[2]Id.

Rules and procedures which are used for a civil action are substantially different from those in a criminal action. First, the parties in a civil action are usually different from those in a criminal action. In a civil action, the plaintiff (the party bringing the action) is usually a private citizen, private corporation, or other party outside of the government and the defendant (the party being sued) is usually a private party, private corporation, or the government. In a criminal action, the party bringing the action is virtually always the government (i.e., U.S. Attorney, State Attorney, local prosecutor), and the defendant is the person charged with the crime. Second, the burden of proof is different in a civil case than in a criminal case. The burden of proof is substantially higher in a criminal case. Third, the rules or procedures are significantly different (i.e., Civil Procedure rules and Criminal Procedure rules). Lastly, the damages sought are usually different.

It should be noted that on certain occasions, a civil suit and criminal prosecution can be brought for the same act. Although these cases would be tried separately, the plaintiff can sue the defendant in civil court for damages while the state could bring a criminal prosecution against the defendant. For example, a fire fighter intentionally strikes you with her fist—you could sue the fire fighter for recovery of your damages which could include such losses as medical costs, lost wages, pain and suffering. Additionally, the state could prosecute the fire fighter for the commission of a crime. Thus, two separate and distinct trials could arise from the same act by the fire fighter. Another example of this is the current O.J. Simpson situation. The State of California has brought criminal charges, and the family of one of the victims has brought a civil action. Separate trials—same issue.

OVERVIEW OF CIVIL ACTIONS

A civil action is brought by one party against another party for the declaration, enforcement, or protection of a prescribed right or redress or prevention of a wrong. In a civil action, the party bringing the action is known as the "plaintiff" and the party being sued is known as the "defendant." The right to sue is considered a fundamental right of all individuals and entities (i.e., fire service corporations), and the suit can proceed provided there is some material but yet unresolved question or issue regarding a right or injury. In most courts, significant deference is provided to parties to bring suit due to the fact that the courts do not want to forestall a party's opportunity to be heard and to have the claim adjudicated.

Fire service organizations should be aware that anyone may bring a lawsuit against any party who causes injury, whether the injury is intentional or negligently inflicted. For civil actions, there need only be some existing party which can be targeted through the processes of law and against whom the court's judgment can be enforced. For example, parties capable of bringing civil actions include individuals, corporations, partnerships, government agencies, associations, foreign governments, Indian tribes, aliens, and even convicted felons. With children, unborn infants, and individuals who are not mentally competent, family members or guardians usually bring the action on behalf of the child or incompetent individual.

The majority of civil actions against fire service organizations stem from the tort liability area. A "tort" is the violation of the personal or private rights of another. In essence, everyone has a right to personal safety and personal liberty. Additionally, everyone has a right to use and enjoy personal property without interference by others. The law protects these basic rights and imposes a legal duty on all others to also respect these rights. When any of these rights are violated, the party violated may sue the violating party in tort.

The law of tort can be categorized by either (1) the nature of the conduct of the party or (2) the nature of the harm to the injured party. Intentional harms to the person include assault, battery, false imprisonment, intentional infliction of emotional distress, negligence, and liability without fault or strict liability. Intentional harms may include injury to the person, damage to tangible property or harms to intangible personal interests.

Under the civil action category of intentional harms to the person, the actions of assault, battery, false imprisonment, and intentional infliction of distress have been used in cases relating to the fire service.

An action for assault is usually used to redress the intentional invasion of a person's interest in freedom from the apprehension of imminent harmful or offensive contact.[3] The components necessary for an action of assault include the following:

1. intent to commit harmful or offensive touching or to create apprehension of the same
2. actual apprehension of imminent harmful or offensive touching

For an action of assault, there is no requirement of an actual touching but words alone usually do not create liability unless, together with other acts or circumstances, the individual is placed in reasonable apprehension of imminent harmful or offensive contact. Additionally, the intent to commit a harmful touching can be transferred from one individual to another. For example, if a individual at a fire scene threatens fire fighter A with dynamite, fire fighter B who is standing next to fire fighter A is placed in apprehension of imminent bodily harm. The individual's intent was to harm fire fighter A but this intent was transferred to fire fighter B.

Like assault, an action in battery is created when the right of the individual to be free from intentional and unpermitted offensive or harmful bodily contact is violated. The components of battery are virtually the same as assault with the exception that an unconsentual physical touching has occurred. In an action for battery, the plaintiff must prove, as part of the *prima facie* case, that he/she did not consent to the touching (i.e., such as a welcome hug or embrace). Consent acquired through duress, where the individual was a minor or lack the mental capacity to consent, or consent was acquired through fraud, misrepresentation, or mistake are usually not effective.

---

[3]Restatement (Second) of Torts § 21.

Of importance to fire service organizations is the potential of assault and battery while rendering medical care beyond first aid. The traditional rule is that a physician or other health care provider could extend an operation only when an emergency exists in the sense that a failure to extend the operation would endanger the patient's life or health, or where a later operation might endanger the life or health of the patient and it is impracticable to obtain the consent of the patient's family.[4] However, some states have adopted a more liberal view. The test usually applied to these circumstances is whether a reasonable person in the position of the patient would grant consent to the extension of the medical care if he/she were able to choose.[5] Additionally, the areas of informed consent and failure to disclose the relevant risks in full are also becoming prominent medical issues sounding within the category of tort law.

The interest protected by the action of false imprisonment is the individual's right to freedom from confinement.[6] The components of an action for false imprisonment include:

1. intent
2. confinement
3. consciousness of the confinement by the individual being confined
4. absence of consent by the person being confined

For false imprisonment to occur, the individual must prove that he/she was restrained from movement without possibility of escape. Merely impeding an individual's progress or confining an individual where a reasonable exit exists usually does not meet this requirement. The majority of the cases in this area include false imprisonment by shopkeepers after alleged shoplifting, unlawful detention in vehicles, or threats of force to prevent movement.

---

[4]*Tabor v. Scobee*, 254 S.W.2d 474 (Ky. 1953). Also see, Restatement (Second) of Torts § 892D.

[5]*Kennedy v. Parrott*, 243 N.C. 355, 90 S.E.2d 754 (1956).

[6]Restatement (Second) of Tort § 35.

The tort of intentional infliction of emotional distress is when an individual," who, by extreme or outrageous conduct, intentionally or recklessly causes severe emotional distress to another . . ."[7] The components of an action for intentional infliction of emotional distress include:

1. intent
2. severe emotional distress (mere embarrassment or humiliation is usually not sufficient)
3. special relationship (i.e., common carriers, innkeepers, public utilities)

Fire service organizations should be aware that many courts have found that where a special relationship exists between the parties, insults of a highly offensive nature are actionable.[8] Additionally, where the defendant directs extreme or outrageous conduct at a person, the defendant may also be liable if the intentional or reckless actions cause severe emotional distress to members of the immediate family present at the time or others who may have witnessed the event (especially when the emotional distress is accompanied by physical harm).[9]

The intentional torts to property include trespass to land, trespass to chattel (personal property), and conversion. The basic components of each of these actions include:

1. possessory interest in the property
2. an intentional entry or taking of the property
3. damage to the property

For fire service organizations, an individual may be privileged in an emergency situation to enter upon or use the property of another to protect himself/herself or other property under a public necessity theory.[10]

---

[7]Id. at § 46.

[8]See, *Slocum v. Food Fair Stores of Florida*, 100 So.2d 396 (Fla. 1958).

[9]Restatement (Second) of Torts § 46.

Closely allied with the individual acting for public necessity is the state's (and thus the fire service's) exercise of the police power to enter property without trespass. In most states, an individual's right to use and enjoy his/her property is subject to the state's right of eminent domain and the lawful exercise of its police powers. Eminent domain involves the taking of property for public benefit (i.e., state's taking and destroying an old building to build a new fire house) while the police power involves the destruction or limitation of the use of property which, under the circumstances, represents a danger to the public safety, public health or public welfare (i.e., destruction of building to serve as a fire break).

The defenses available for the above intentional torts fall within the categories of privilege and immunity. These defenses usually include:

1. privilege
   a. self defense—use of force necessary to reasonably repel an attack. Usually there is no obligation to retreat.
   b. defense of others—use of same force as the third person may use
   c. defense of property—no deadly force permitted
   d. recapture of property—demand for return of property usually required and no use of deadly force
   e. merchant's privilege—usually statutory—permits appropriate detention of shoplifters
   f. discipline of children—reasonable force for proper control, training.
2. immunities
   a. interspousal immunity
   b. parent and child immunities
   c. charitable immunities
   d. governmental immunities
      • federal government immunities—must consent

---

[10]See, *Ploff v. Putnam,* 81 Vt. 471, 71 A. 188 (1908).

to be sued under the Federal Tort Claim Act.[11]
- state government immunities—usually proscribed by statute in the state, municipal government immunities, limited immunity (Chapter IV)
- public officers immunities (Chapter IV)

The tort law category under which a substantial number of cases have been brought against fire service organizations is that of negligence. The components of a negligence action include:

- duty—whether the fire service organization owed a duty to the plaintiff and, if so, the standard of care the fire service organization owed the plaintiff under the circumstances
- breach—whether the fire service organization or individual fire fighter, through their conduct, violated that duty of care
- harm and causation—whether the fire service organization's conduct factually or proximately brings about the actual harm to the plaintiff
- damages—the harm done

The defenses to a negligence action usually fall within the categories of substantive defenses (i.e., contributory negligence or assumption of the risk) or procedural defenses (i.e., immunity from prosecution).

Under the first component of duty, the general rule is that the law does not impose the duty to act in any way upon individuals. A duty can be created by law, by the relationship between the parties, or by one party creating the harm to another. In the case of most fire service organizations, the duty to act has been created by statute or law.

---

[11]Title 28, U.S.C.A. §1346.

For fire fighters away from the job, if an individual has no duty to render aid or assistance undertakes to render aid or assistance, he/she is required to exercise reasonable care if assistance is voluntarily provided.[12] Many states have enacted Good Samaritan statutes which afford a degree of protection against suit for rendering aid. Fire fighters should note that several states have additionally enacted statutes making the Good Samaritan rule applicable to doctors, nurses, EMTs and other health care professionals who render medical assistance at the scene of an accident. These laws usually absolve the health care professional from liability for their ordinary negligence in rendering emergency first aid. Additionally, the standard with which individuals are confronted in an emergency situation is usually different (i.e.,the individual is not required to act as if he/she had adequate time to weigh alternatives and decide the most reasonable course of action).

In a rescue situation, if the individual initiates a rescue and thus prevents another from initiating a rescue, the rescuer usuallyy has created a duty to complete the rescue. Once a rescue is initiated and thus a duty is created, the rescuer must complete the rescue or possess appropriate reasons for discontinuing the rescue attempt.

The degree of knowledge and professional qualifications are within the category of duty. Children below the age of seven are usually presumed incapable of negligence. Between the ages of seven and fourteen, some courts apply the rebuttable presumption that the child is incapable of negligence. Children performing adult activities, such as operating a snow mobile, airplane, or boat, are usually held to an adult standard. An individual's mental deficiencies or capabilities are usually not considered. However, physical infirmities are often considered in determining the component of duty. Professionals, such as physicians, are usually held to a professional standard of care.

Most states treat fire fighters as licensees for the purposes of entering private property.[13] This rule is based on the premise that

---

[12]See, *Zelenko v. Gimbel* Bros. 158 Misc. 904, 287 N.Y.S. 134 (1935).

fire fighters are likely to enter the property at unforeseeable times and in unforeseeable ways. Some states, however, have held that fire fighters are invitees when they enter a property under the same circumstances as other members of the public.[14] It should be noted that several cases have held that fire fighters are to be treated in all respects as business invitees, and that the unusual aspects of the time and location of the entry on the property merely goes to the issue of foreseeability.[15]

Fire service organizations should be aware that virtually all states have established the special rule for infant and children trespassers known as the "attractive nuisance doctrine."[16] This doctrine provides that the possessor of land (such as fire stations) is subject to liability for physical harm to child tresspassers caused by an artificial condition on the land if the following conditions are present:

1. The possessor knows or has reason to know that children are likely to trespass in the location of the danger the possessor knows or has reason to know that the condition will involve an unreasonable risk of death or serious bodily harm to such children.
2. The children, because of their youth, do not discover the condition or realize the risk involved.
3. The utility to the possessor of maintaining the condition and the burden of eliminating the danger are slight as compared with the risk to children involved.
4. The possessor fails to exercise reasonable care to eliminate the danger or otherwise to protect the children.

---

[13]See, *Mulcrone v. Wagner*, 212 Minn. 478, 4 N.W.2d 97 (1942). Entry on to private property can be categorized as trespasser, invitee, or licensee. Generally, the landowner owes no duty to a trespasser, owes a duty of ordinary care to an invitee, and owes a duty of reasonable care to the licensee).

[14]*Meiers v. Fred Koch Brewery*, 229 N.Y. 10, 127 N.E. 491 (1920).

[15]*Dini v. Naiditch*, 20 Ill.2d 406, 170 N.E.2d 881 (1960); Cameron v. Abatiell, 127 Vt. 111, 241 A.2d 310 (1968).

[16]Restatement (Second) of Tort § 339.

Given the bells, alarms, sirens, activity, and other items which tend to attract small children to fire stations, this is an especially important area of concern for fire service organizations. Fire service organizations should carefully evaluate their fire stations to ensure that appropriate precautions are taken to protect small children who may be playing in and around the area.

The second component necessary for a negligence action is a breach of the duty of care. The burden of proving that the defendant violated the duty owed is placed solely on the plaintiff. The three basic methods of proving a breach of the duty are:

1. *res ipsa loquitur*—"the thing speaks for itself"—(i.e., circumstantial evidence)
2. violation of a statute
3. direct evidence of negligence

In most cases involving a fire service organization, the breach component is easily achieved by the plaintiff because the duty is created by law, statute or ordinance. This is often know as "negligence *per se*." A breach can result from nonfeasance (i.e., the failure to perform a required duty such as showing up to a fire); by misfeasance (improper performance of a required duty such as not using water on a structure fire); or by malfeasance (e.g., deliberate violation of a legal duty, such as permitting the house of a person suing the fire department to burn without response). A breach can be by commission or omission.

The third component in a negligence action is proof of actual or proximate causation and harm to the plaintiff by the defendant. The plaintiff must prove that he/she has suffered a harm to themselves or their property. In most situations, there is usually no recovery for pure economic harm such as lost profits or wages.

In most cases, the causation component relates to the proximate cause of the harm, such as whether the defendant's negligent act was sufficient to be considered the legal cause of the harm. The court must determine whether the defendant's act or conduct was actually connected with the plaintiff's harm. If the plaintiff is unable to prove the harm was related to the defendant's negligence, the plaintiff is not entitled to recover for the harm.

The two basic tests used for this determination is the "but for" test and "substantial factors" tests. In most cases, the "but for" test is utilized (also known as the *sine qua non* rule). Under this test, the court determines where the harm to the plaintiff would have occurred "but for" the negligent acts of the defendant. For example, the attached garage on the house would not have burned but for the negligence of the fire department in failing to show up to the fire.

The "substantial factors" rule is used in special cases where, under the "but for" test, the defendant would escape liability because of other contributing causes. For example, two individuals fire shotguns in the general direction of the plaintiff but only one hits the plaintiff. Both defendants acted negligently, although independently, and it is impossible to determine which defendant caused the harm to the plaintiff.

Where two or more defendants commit negligent acts but it is impossible to show which injuries were caused by which defendant, most courts, under the theory of joint and several liability, hold all of the wrongdoers liable for all of the plaintiff's injuries.[17] For example, a fire service vehicle is struck by two other vehicles simultaneously at a traffic light.

Assuming that actual causation exists between the defendant's conduct and the harm to the plaintiff, no recovery is permitted unless the defendant's negligent act was the proximate cause of the harm. Proximate cause (also known as legal cause) can be delineated into two broad categories, namely: (1) harm within the risk, and (2) persons within the risk. In most states, the defendant is liable only if the harm was a foreseeable consequence of the unreasonable act. The minority position adopted by some states is that the defendant is liable for any harm which follows in an unbroken

---

[17]See, *Maddux v. Donaldson,* 363 Mich. 425, 108 N.W.2d 33 (1961).

sequence from the negligent act of the defendant. It should be noted that these tests are viewing the harm in hindsight after the incident. Negligent intervening acts or intentional intervening acts of a third party have an effect on this determination.

Of importance to fire fighters is the use of the rescue doctrine in determining the foreseeability of the risk. Under this doctrine, a defendant who negligently places a victim in a perilous position owes a duty to the person coming to the aid of the victim. Thus, if the rescuer is injured in making a rescue, the defendant will be liable to the rescuer as long as the rescuer exercises reasonable care.[18]

The last component in a negligence action is damages. Damages usually consist of compensatory damages (medical costs, lost wages, future losses, pain and suffering) and, in some circumstances, punitive damages (damages provided to deter future actions or to set an example).

Under common law, there was no recovery of damages for wrongful death. In essence, the potential lawsuit died with the victim. Today, fire service organizations should be aware that most states have adopted "wrongful death" statutes which permit the families or beneficiaries to sue on behalf of the deceased victim.

Another important area for fire service organizations is the emerging area of prenatal injuries and death. In most states, if a child is born alive, the child may recover from a negligent defendant for injuries sustained while in utero. If the child dies after being born alive, an action for wrongful death is usually permitted.

In the majority of the states, if a fetus dies in utero as a result of the defendant's negligence, providing the fetus was viable at the time of death, a wrongful death action is permitted. Conversely, most courts do not permit a claim for damages based on a "wrongful birth" theory.[19]

---

[18]See, *Wagner v. International Railway,* 232 N.Y. 2176, 133 N.E. 437 (1921).

[19]See, *Glutamine v. Cosgrove,* 49 N.J. 22, 227 A.2d 689, 22 A.L.R.3d 1411 (1967). Also see, Turpin v. Sortini, 31 Cal.2d 220, 643 P.2d 954 (1982)(Special medical expenses in the past and future may be permitted).

For damages to be recovered for mental distress brought about by the negligent conduct of the defendant, most states require that the plaintiff prove physical harm.[20] The Restatement (Second) of Torts § 436A takes the position that recovery may not be had if the defendant's negligence causes only mental distress to the plaintiff, without any physical harm or other compensable damage being produced by mental distress. Transitory, nonrecurring physical phenomena, harmless in themselves, such as dizziness, vomiting, and the like, do not make the defendant liable where such phenomena are in themselves inconsequential and do not amount to any substantial bodily harm. Additionally, many states require that the plaintiff's mental distress must be within the "zone of danger" in which the plaintiff actually observed or was placed in physical danger.[21]

The defenses which may be available to a fire service organization to combat a negligence action include the substantive defenses of contributory negligence and assumption of the risk. The procedural defenses may include immunity among other defenses. (See defenses to intentional torts.)

Most states have adopted some form of comparative negligence rather than the contributory negligence theory. Under the comparative negligence theory, the plaintiff is permitted to recover his/her damages reduced by the percentage of negligence attributable to the plaintiff. Under a pure contributory negligence theory, if the plaintiff is somewhat negligent, this negligence could completely bar recovery of damages. The adoption of some form of comparative negligence by states is usually done by a change in the state statutes.

Closely aligned with the theory of contributory negligence is the defense of assumption of the risk (also known as *volent non fit injuria*). The defense of assumption of the risk requires proof that the plaintiff knowingly entered into, or stayed in, a position of danger. In many comparative negligence jurisdictions, the assumption

---

[20]See, *Bosley v. Ansrews,* 393 Pa. 161, 142 A.2d 263 (1958).

[21]See, *Sadler v. Cross,* 295 N.W.2d 552 (Minn. 1990).

of the risk defense has been abolished as a separate defense and merged into the comparative negligence analysis. Fire service organizations should be aware that this defense may be available where the employer fails to provide a reasonably safe working environment.[22]

The nature of the defendant's liability is often at issue in cases involving a fire service organization. Where two or more defendants bring about the harm and it is impossible to separate the portions of harm, courts often find both parties jointly liable. The parties are known as "joint tortfeasors." In these circumstances, the tortfeasors may sue third parties who may have some liability for contribution for the damages.

Of importance to fire service officers and fire fighters are the doctrines of indemnification. In contribution, all defendants share the financial burden jointly among all tortfeasors. Under indemnification, the fire service organization accepts responsibility for all damages resulting from the acts of its officers or fire fighters in order to protect the fire fighters from personal liability. Indemnification clauses are often included in employment contracts or other documents setting forth the conditions of employment. For example, if a fire service officer, working in the scope of his/her employment, is sued for negligence, under the doctrine of indemnification the fire service organization would accept responsibility for the costs incurred by the officer in defending against the action and pay any/all damages which were attributed to the officer's actions.

Closely aligned with the doctrine of indemnification is vicarious liability. A fire service organization may be liable under some circumstances for the actions of the officers and fire fighters because of the relationship between the fire service organization and the fire fighter. Under the common law principal of Master and Servant, the master is vicariously liable for the torts of his/her servants who are acting within the scope of their employment. In most fire service

---

[22]See, *Siragusa v. Swedish Hospital,* 60 Wash.2d 310, 373 P.2d 767 (1962).

organizations, the fire service organization is the employer, and thus the master, and most fire fighters are employees, and thus servants under this principle. If the fire service organization has the right of control or actually controls the actions of the fire fighters, vicarious liability may attach. This principle may also attach to agents of the fire service organization, borrowed fire fighters, and others working for the fire service organization so long as the fire service organization exercised control over the activities and actions of the individual. Fire fighters should be aware that in many states parents can be held vicariously liable for the actions of their children.[23]

The defense of immunity is normally based upon public policy considerations and is designed to protect the fire service organization despite the fact that a negligent act may have been committed. There are several types of immunity provided under the law, including interspousal immunity, parent-child immunity, charitable immunity and a variety of types of governmental immunity. The varying degrees of governmental immunity exist at the federal, state, and municipal levels of government service. The governmental immunity area is one area in which fire service organizations may acquire protection in certain circumstances.

The federal government usually may not be sued in tort except to the extent that it has consented to suit by the enactment of the Federal Tort Claims Act (28 U.S.C.A.). Under this Act, the federal government will be liable where "the United States, if a private person, would be liable to the claimant in accordance with the laws of the place where the act or omission occurred." (28 U.S.C.A. §1346). The Federal Tort Claims Act specifically states that the United States should not be liable for assault, battery, false imprisonment, and a variety of other acts.[24] The federal government is not liable under the act for performances or non performance of a *discretionary* function or duty on the part of one of its agencies or

---

[23]See, *Cardwell v. Zaher*, 344 Mass. 590, 183 N.E.2d 706 (1962)(parents knew or should have known of the child's propensity for such conduct).

[24]28 U.S.C.A. § 1680 (h)

employees.[25] A distinction needs to be drawn between the functions at the planning level as opposed to the functions at the operations level and liability under the Federal Tort Claims Act exists only as to the functions of the operations level.[26]

On a state level, the doctrine of sovereign immunity has long been established. Thus, a fire service organization working as a state governmental function may be sued for torts committed by their employees and agencies only to the extent that they have, by statute or law, consented to such suits. All states, in varying degrees, have consented to negligence actions, and normally there is an established mechanism to address these claims. The Restatement of Torts § 895 (b) takes the position that while a state is not liable to suit for its torts without its consent, such consent need not come in the form of legislation. State immunity may be abrogated or restricted through court decisions and legislature has the prerogative to outline the preferred form of the action. Even where consent is given, the Restatement takes the position that it should be limited consent and the state governmental group should be immune from liability for its actions and omissions which constitute an exercise of judicial or legislative functions or the exercise of administrative functions involving basic policy decisions.

Municipal corporations have never enjoyed complete immunity from tort liability as with the federal and state governmental entities. This is because the municipality performs functions which are both governmental in nature and proprietary in nature. Traditionally, municipalities, city governments, and fire service organizations have been held liable for torts committed by their employees in the performance of proprietary functions but have not been held liable where a tort arises out of the performance of a governmental function.

The doctrine of sovereign immunity has been dissipating rapidly in recent years. In the past, fire service organizations could rely

---

[25]28 U.S.C.A. §2690 (a).
[26]*Dalehite v. U.S.*, 346 U.S. 15, 73 S. Ct. 956, 97 L. Ed. 1427 (1953).

extensively on the doctrine of sovereign immunity to avoid negligence actions being brought against their organization. Although this doctrine is still viable in many jurisdictions, fire service organizations can no longer rely solely on the sovereign immunity doctrine to exempt them from legal action.

As a general rule, fire service organizations are not responsible for the acts of independent contractors. This is based on the idea that fire service organizations would not have the right to control activities of independent contractors. However, case law and statutes have recognized certain situations in which an employer or organization cannot delegate the specific duty.[27] In most circumstances, however, where the fire service organization does not exercise control over the contractor, the truly independent contractor accepts responsibility for his/her own acts. Fire service organizations are cautioned to ensure that appropriate protections are included in all written contracts with independent contractors performing work for the fire service organization. For example, if the fire service organization was building a new fire house, the fire service could contract the entire job of building the structure to an independent contractor. Usually by written contract, the fire service would provide the specifications but the independent contractor would accept responsibility for the job site and the workers' safety. However, if the fire service organization, in an attempt to cut costs, assigned an officer to oversee the project acting in the capacity of the general contractor and subcontracted all work to other contractors, the fire service organization, because of their exercise of control, may vicariously be accepting responsibility for the safety of the workers at the job site and other liabilities involved in the construction of this fire house.

One last area that fire service organizations and fire fighters should be aware of is the negligent entrusting of a motor vehicle. If a fire service organization entrusts a vehicle to an unfit person, the

---

[27]See, *City of Baltimore v. Lleonard,* 129 Md. 621, 99 891 (1917) (city could not delegate reasonable repair of streets to contractor).

fire service may be held liable for resulting injury to others if the fire service organization knew of the operator's unfitness.

For example, if a fire fighter has been drinking at the annual Christmas party and then takes a fire service vehicle to respond to a call and ultimately has an accident in the vehicle to the officer with the knowledge that he/whe was under the influence, the fire service organization may have negligently entrusted this vehicle and thus may be responsible for the resulting injuries.

Fire service organizations should be aware that the above are but a few of the potential theories in which an action can be brought by or against a fire service organization. Other potential actions may lie in the areas of product liability, strict liability, and harm to economic interests such as deceit, negligent misrepresentation, interference with contractual relationships, defamation, malicious prosecution, and invasion of privacy.

Common Deficiencies Leading to Vicarious Liability[28]

Negligent Appointment
- Failure to check a person's background and qualifications prior to employment or membership, or assignment to a task.

Negligent Hiring
- Hiring an individual with known or easily found deficiencies.

Negligent Retention
- Keeping a person in a job or position which he has proved he cannot do, once you found out he can't do it.

Negligent Assignment
- Assignment of a person to a job which he cannot do, for which he is not qualified, or is untrained.

Negligent Entrustment
- Ordering or permitting a person to use some piece of equip-

[28]Taken from *The Seven Deadly Sins of Vicarious Liability*, by Normal L. Lawson, Jr.

ment or device for which he is not trained, or does not know how to use, or cannot use properly.

Failure to Train

- Not providing training for personnel according to their unsupervised manner at any time or under any condition where supervision is needed.

Failure to Direct

- Failure to have rules and regulations, standard operating procedures, instructions, guidelines, and the enforcement of same which relate to the operation of the department and the conduct of its activities.

COURT PROCEDURES IN CIVIL ACTIONS

The right to sue another for harm and damages is a fundamental right of all individuals and entities in the United States. A civil action through our judicial system is a method to resolve conflicts between the parties with regards to material issues such as the extent of injury or the amount of damages. In essence, anyone may bring a civil lawsuit against another person who causes injury, whether the injury was negligently inflicted or intentionally inflicted.

There are, however, numerous limitations on the right to bring a civil action and, as noted above, which court to bring the action before. The most frequently used limitation which can bar a civil action from taking place is the passing of the time limitations specified by statute. This time limitation is often referred to as the "statute of limitations." For example, if a wrongful discharge action has a 1 year statute of limitations and if the action is brought on the 366th day, the action would be barred by the statute of limitations.

In addition to the statute of limitations defense, other limitations are usually placed upon the action as to the jurisdiction and venue.

These limitations can include federal or state designations, monetary requirements, and numerous other limitations. Failure to comply with all provisions and restrictions can result in the court's dismissal of the action.

Prior to the filing of a civil action, there are numerous considerations which must be evaluated. The rules and requirements which must be followed are usually included in the state or federal statutes and the Rules of Civil Procedure. First, the appropriate court must be selected. As noted above, each and every rule and restriction must be analyzed to ensure the "right" court is selected. Second, the "key" which starts the lawsuit is the filing and service of a complaint against the other party. The complaint (sometimes called a "pleading") is usually a plainly written notice to the other party as to why they are a party, why damages have been incurred, which court will hear the case, and what damages have been incurred. A complaint is often attached to a summons informing the party to appear and defend the action under penalty of default (see Exhibit 3.1).

Following the service of the complaint, the defendant must respond within a specified number of days. The defendant may challenge the complaint by a motion to dismiss, a motion to dismiss for failure to state a claim (also known as a demurrer), or other challenges. If these motions are denied, the defendant must file an "Answer" to the complaint. Importantly, affirmative defenses usually must be pleaded in the Answer. There may be an additional response to the Answer by the plaintiff which is known as a "Reply." Additionally, the action can be expended through including other parties (known as impleading) or the defendant may bring a claim against the plaintiff (known as a counter claim).

The method of acquiring information from the opponent is known as "discovery." Discovery may include written interrogatories (a series of questions), depositions (questioning of witnesses under oath), Orders for production of documents, and requests for admissions among other methods.

Prior to trial, the defendant usually files a Motion for Summary judgment. In essence, since the court could not dismiss a complaint

Exhibit 3.1 Sample Complaint

IN THE CIRCUIT COURT OF XXXXXXX COUNTY,

COMMONWEALTH OF KENTUCKY

CIVIL ACTION NO._____

Mr. John Doe, Plaintiff,   COMPLAINT
vs.
X County Fire Department, Defendant.

COME THE PLAINTIFF, John Doe, a citizen of Kentucky, for his claims and causes of actions against the defendant, states as follows:

JURISDICTION AND VENUE

1. The plaintiff and defendants are all citizens of the Commonwealth of Kentucky.
2. The matter in controversy exceeds the sum of four thousand dollars ($4,000.00), exclusive of interest and costs.

CLAIM AGAINST DEFENDANTS

3. That on February 10, 1994, the defendant fire department, through its employee, Larry Collins, did negligently operate a fire pumper which collided with the plaintiff's vehicle at the intersection of Main Street and Third Street, Richmond, Kentucky.
4. That the plaintiff did, as a result of the negligent operation of the fire pumper, suffer permanent physical and mental injuries, property damage to the vehicle, and additional damages.

WHEREFORE, the plaintiff demands judgment against the defendant in the amount of x thousand dollars, other damages as deemed necessary, pre-judgment interest, post-judgment interest, cost expended herein, and hereby requests a trial by jury.

_____
ATTORNEY FOR PLAINTIFF
Address & Telephone #

---

so long as the complaint was based on a good faith belief and basis for the complaint, if the discovery phase does not produce adequate evidence to support the plaintiff's contention, the court may summarily dismiss the action on this motion. If the action survives the Motion for Summary judgment, the case is usually set for trial.

It should be noted that the vast majority of civil cases are settled prior to trial. However, if the parties cannot agree to a settlement and a trial is required, typically one or the other parties file a Note of Issue at which time the case is provided a number and placed on the trial calendar. In most states, the trial calendars have become extremely long and a case may have to wait a year or longer before it is called to trial.

At trial, the components of the trial include:

1. selection of a jury (if applicable)
2. questioning of the jurors by each side and the removal of jurors for cause or by preemptory challenge
3. opening statements by each side (usually plaintiff goes first)
4. direct examination of witnesses (by plaintiff)
5. cross examination of witnesses (by defendant)
6. upon closing of the plaintiff's case in chief, the defendant usually makes a motion for directed verdict where the judge may remove the case from the jury if the plaintiff did not prove his/her case
7. direct examination (by defendant)
8. cross examination (by plaintiff)
9. closing statements by both sides
10. deliberation and decision by jury
11. post-trial motions (i.e., motion notwithstanding the verdict, new trial)
12. appeal by defendant or enforcement of the judgment by the defendant

It should be noted that the remedies for a civil suit which the jury or judge is permitted to provide are usually set by statute or part of the substantive law of the jurisdiction. Not every remedy is available in all actions.

OVERVIEW IN CRIMINAL ACTIONS

Criminal actions are brought against an individual or other entity who has harmed the interests of our society. Criminal laws can be divided into five basic categories: (1) homicide, (2) nonhomicide crimes against a person, (3) crimes against property interests, (4) crimes against habitation, and (5) inchoate crimes.

Fire service organizations and individual fire fighters should be aware that they are not immune from committing a criminal act or being victimized by a criminal act. As noted in the Arson section, a fire fighter can be susceptible to the lure of such criminal acts because of their specialized training and expertise and the access which employment within the fire service can provide to an individual inclined to such criminal activities.

Homicide is the death of a human being caused by another. To have a homicide, the death must be of a born alive human being by another human being and the human act must be the sole or concurrent substantial "but for" proximate cause of the death. In most states, homicide is categorized under various levels including murder and degrees thereof and manslaughter, voluntary and involuntary.

Murder is usually defined as a criminal homicide committed with malice aforethought. Malice aforethought is a technical term which encompasses the intent to kill (i.e., expressed malice, willful and wanton disregard, in the commission of a crime, and other technical definitions). For first degree murder, premeditation and deliberation by the person committing the murder or attempting commission of certain dangerous felonies (i.e., felony-murder rule) are required. First degree murder can also include such situations as torture or extreme cruelty prior to murder, use of poison in the murder, and lying in wait prior to the murder. Second degree murder is usually any murder which does not meet the definitions set forth for first degree murder, i.e., without premeditation and deliberation.

Voluntary manslaughter is the intentional homicide under extenuating circumstances which mitigate but do not justify or excuse the killing. Usually, the element of malice aforethought which is required for murder is not present. The most common types of voluntary manslaughter involve the intentional killing while in the heat of passion (absent an adequate "cooling off" period) caused by adequate provocation (i.e., loss of control). Voluntary manslaughter situations providing adequate provocation include cases of a spouse observing his or her spouse in the act of committing adultery, an attempted battery if the result would arouse the passions of a reasonable man, and mutual combat situations. It should be noted that words alone, no matter how abusive, do not constitute adequate provocation.[29] Other forms of voluntary manslaughter can include assisting a suicide, where the defense of self-protection or defense of others is imperfect, or the defendant has a diminished capacity. (It should be noted that voluntary drunkenness is usually not sufficient to reduce murder to manslaughter.)

Involuntary manslaughter is an unintentional homicide without malice which is neither justified nor excused. Involuntary homicide can include criminal negligence (such as a person cleaning a loaded gun and the gun accidentally discharges killing a bystander) and a killing during the commission of a misdemeanor (misdemeanor-manslaughter rule).

Homicide can be justifiable where the homicide was authorized by law. These situations include self-defense, defense of others, defense of one's dwelling, killing under public authority (i.e., police, FBI) or during an arrest, escape, or prevention of a crime. In the category of nonhomicide crimes against persons, the basic offenses of assault and battery are present in addition to robbery, extortion, kidnapping, and rape. The basic requirements of criminal assault and battery are the same as the intentional torts of assault and battery. The common law felony of robbery is larceny from the person through violence or intimidation. To prove robbery, there must be a taking from a person by force or intimidation and the taking and the force must be coincidental to the use fo force. At common law, extortion was the corrupt demanding or receiving by a public official of a fee for services which should have been performed gratuitously. Under the Model Penal Code § 223.4, a person is guilty of theft by extortion if he purposely obtains property of another by threatening to:(1) inflict bodily harm . . . ,(2) accuse another of a criminal offense, (3) exposes any secret tending to

---

[29]See, *Freddo v. State*, 127 Tenn. 376, 155 S.W. 170 (1913).

subject any person to hatred, contempt or ridicule or impair his credit or business reputation, (4) take or withhold action as an official, or cause an official to take or withhold action, (5) bring about or continue a strike, boycott or other collective unofficial action . . . ,(6) testify or provide with respect to another's legal claim or defense, or (7) inflict any other harm which would not benefit the actor.

Kidnapping is an aggravated form of false imprisonment involving any asportation or secrecy. Additionally, many states have adopted statutory variations of the common law false imprisonment which cover such areas as kidnapping for ransom and child abduction.

Rape is unlawful sexual intercourse with a female without her consent. Rape can be by force or through fraud or inducements. Statutory rape usually involves unlawful intercourse with a female below the age of consent (usually age sixteen) and consent is usually not a defense.

The category of crimes against property include larceny, embezzlement, obtaining property by false pretenses, and receiving stolen goods. Although most of these crimes are self-explanatory, the categories of larceny and embezzlement are often confused. The basic difference is that larceny would be, for example, when a fire fighter takes a tool from the firehouse and keeps the tool, whereas embezzlement would be the fire chief giving the fire fighter $50.00 to buy a tool and the fire fighter buys the tool from the dealer and then keeps the tool.

Of particular importance to fire service personnel is the category of crimes against habitation. These crimes include burglary and arson. At common law, burglary is the breaking and entering of the dwelling of another in the nighttime with the specific intent to commit a felony therein. Most states have broadened this category to eliminate the breaking and the nighttime requirement and added such other categories as cars, and all buildings.

Arson is a common law crime which has been extensively expanded by state legislatures. At common law, arson was defined as the malicious burning of the dwelling of another. Common law arson required that the individual had the intent to burn the dwelling, actually set the fire, and the building was required to be another person's home. Most modern statutes have expended the scope to include all buildings or other personal property, such as automobiles, and provided for separate laws when an individual burns his or her own property for insurance proceeds.

Inchoate crimes include attempts to commit a crime, solicitation, and conspiracy to commit a crime. Under these laws, the individual may not be required to actually have completed the crime and must withdraw within a specified time period in order to not accept responsibility. These crimes are often included with other crimes or used to compound the offense.

COURT PROCEDURES IN CRIMINAL ACTIONS

The rules and limitations on criminal actions are significantly more stringent than civil actions. By statute and case law, the federal government and virtually all states have developed specific rules (such as the Rules of Criminal Procedure) and limitations (such as the 70 day requirement for a speedy trial). It should be noted that the burden of proof in a criminal case is substantially higher and the burden is placed solely on the state or federal government. Remember, the stakes are much higher in a criminal case because the ultimate outcome can include the removal of an individual's personal freedom.

In the criminal area, the case usually starts with a grand jury determination as to whether the prosecutor has sufficient evidence to charge an individual. If sufficient evidence is present, a warrant for arrest is issued and the individual charged is arrested. The first interaction with the court is when the individual is arraigned (i.e., charges are read to the defendant). At arraignment, the defendant is provided an attorney if he/she cannot afford one and bond is determined. The defendant may be released on bond or may remain in custody pending trial.

At this stage, the defendant is entitled to view all evidence against him/her and motions are usually made to exclude evidence from trial. Additionally, other motions are usually made regarding virtually every aspect of the prosecutors case. The court normally decides on these motions at a "motion hearing."

In many criminal cases, the parties agree (known as a "plea agreement") to a crime or sentence if the defendant agrees to plead guilty to the offense or some other offense, usually lower. The court must approve the plea agreement.

If the criminal case "goes to trial," the components are virtually identical to that of a civil trial. The differences include the fact that the plaintiff would now be the prosecutor and the person being charged is the defendant, the rules of criminal procedure are used, and the jury or judge determines the sentence rather than a monetary amount. Appeals from criminal cases are much more frequent and the individual is often incarcerated while the appeal is pending.

The amount time to be served and/or the amount of monetary fines which can be assessed are usually determined by statute or other laws and are usually set in accordance with the type of offense. For federal offenses, the Federal Sentencing Guidelines are used which provide categorization of the crime, increases in sentence for criminal history, and other increases and decreases for various other factors. For example, voluntary manslaughter carries a base level offense of 25. If this individual is convicted of a voluntary manslaughter in a federal court and, among other factors the individual had no prior convictions, the sentence would be in Criminal History Category I with a offense level of 25 which provides the court may sentence the individual to between 57 and 71 months of incarceration. However, if another individual committed the same offense but had prior convictions sufficient to move him from a Criminal History Category I to Category VI, the sentence would be between 110–137 months.

(Note: The Federal Sentencing Guidelines are currently receiving substantial criticism by many commentators.)

Sentencing Table (in months of imprisonment)

Offense Level	Criminal History (Criminal History Points)					
	I (0 or 1)	II (2 or 3)	III (4, 5, 6)	IV (7, 8, 9)	V (10, 11, 12)	VI (13 or more)
1	0–6	0–6	0–6	0–6	0–6	0–6
2	0–6	0–6	0–6	0–6	0–6	0–6
3	0–6	0–6	0–6	0–6	2–8	3–9
4	0–6	0–6	0–6	2–8	4–10	6–12
5	0–6	0–6	1–7	4–10	6–12	9–16
6	0–6	1–7	2–8	6–12	9–15	12–18
7	1–7	2–8	4–10	8–14	12–18	15–21
8	2–8	4–10	6–12	10–16	15–21	18–24
9	4–10	6–12	8–14	12–18	18–24	21–27
10	6–12	8–14	10–16	15–21	21–27	24–30
11	8–14	10–16	12–18	18–24	24–30	27–33
12	10–16	12–18	15–21	21–27	27–33	30–37
13	12–18	15–21	18–24	24–30	30–37	33–41
14	15–21	18–24	21–27	27–33	33–41	37–46
15	18–24	21–27	24–30	30–37	37–46	41–51
16	21–27	24–30	27–33	33–41	41–51	46–57
17	24–30	27–33	30–37	37–46	46–57	51–63
18	27–33	30–37	33–41	41–51	51–63	57–71
19	30–37	33–41	37–46	46–57	57–71	63–78
20	33–41	37–46	41–51	51–63	63–78	70–87
21	37–46	41–51	46–57	57–71	70–87	77–96
22	41–51	46–57	51–63	63–78	77–96	84–105
23	46–57	51–63	57–71	70–87	84–105	92–115
24	51–63	57–71	63–78	77–96	92–115	100–125
25	57–71	63–78	70–87	84–105	100–125	110–137
26	63–78	70–87	78–97	92–115	110–137	120–150
27	70–87	77–96	87–108	100–125	120–150	130–162
28	78–97	87–108	97–121	110–137	130–162	140–175
29	87–108	97–121	108–135	121–151	140–175	151–188
30	97–121	108–135	121–151	136–168	151–188	168–210
31	108–135	121–151	135–168	151–188	168–210	188–235
32	121–151	135–168	151–188	168–210	188–235	210–262
33	135–168	151–188	168–210	188–235	210–262	235–293
34	151–188	168–210	188–235	210–262	235–293	262–327
35	168–210	188–235	210–262	235–293	262–327	292–365
36	188–235	210–262	235–293	262–327	292–365	324–405
37	210–262	235–293	262–327	292–365	324–405	360–life
38	235–293	262–327	292–365	324–405	360–life	360–life
39	262–327	292–365	324–405	360–life	360–life	360–life
40	292–365	324–405	360–life	360–life	360–life	360–life
41	324–405	360–life	360–life	360–life	360–life	360–life
42	360–life	360–life	360–life	360–life	360–life	360–life
43	life	life	life	life	life	life

4

Civil Liability with Fire Departments

When the 30-year-old lawyer died, he said to St. Peter, "How can
you do this to me—a heart attack at my age? I'm only 30." Replied
St. Peter, "When we looked at your total hours billed, we figured you
were 95."

Anon.

HISTORICAL UNDERPINNINGS

In the past, fire service organizations were often placed within the
category of a governmental agency and thus, under the concept
"the king can do no wrong," fire service organizations were rarely,
if ever, found liable under a tort liability theory. Through the years,
this well established principle of law that fire service organizations
were considered to be a governmental undertaking, i.e., an undertaking that could only be performed by an agency of the government, has began to erode. Under the general rule, fire service organizations owe a duty to the community as a whole and thus were
not designed to protect an individual's interest. The duty owed by
the fire department was considered to be limited in nature and
restricted to protecting the community in total. In the concept of
immunity, fire service organizations, as a governmental undertaking, were virtually immune to any type of tort liability for loses of
individuals. This age old concept that a fire service organization
owed a duty to all but owed a duty to no one in particular has
received extensive criticism and is slowly deteriorating. Through
the years, courts have established many exceptions to this general
rule of non-liability. For example, many courts have established

exceptions to the rule concerning construction and maintenance of fire department facilities, repair of fire alarm systems, and operation of fire service vehicles. The basic premise in which fire service organizations can determine the extent of their immunity from tort liability usually focuses on the exercise of professional judgment or lack thereof. In general, the court decisions regarding fire service organization liability or immunity tend to fall into the following basic concept that when fire service organization activity is within the area of proprietary functions (i.e., no professional experience is required), no immunity exists for the fire service organization. Thus, the fire service organization and the municipality and the other governmental agency may be liable under the same tort liability theory as a private corporation.

For activities that are within the realm of a governmental function, i.e., require the use of a professional judgment of fire service personnel, the fire service organization, municipalities, or any other government agency could be responsible under the doctrine of *respondeat superior* (i.e., the fire service organization was responsible for the wrongful acts of its fire fighters). This theory of tort liability has been used when a fire fighter or other agent of the fire service organization was in privity with the fire service organization or where the fire fighter or agent detrimentally relied upon the word or actions of the fire service organization. Immunity for the fire service organization when the activities of the fire service organization fell within the categories of the judicial, quasi judicial, legislative, or quasi legislative character. If the activity fell within the exercise of governmental judgment, the fire service organization, municipality, or other governmental agency was usually immune to tort liability.

CITY AND MUNICIPAL LIABILITY

As noted above, municipal immunity is based on the theory that "the king can do no wrong." Through varying exceptions that have been carved out of this general rule and the concept of immunity has been transferred to the fire service organizations. The issue of tort liability for fire service organization, municipalities, and other governmental agencies can be a confusing situation. The confusion stems from a clash between theories of law which are fundamental to the subject. Thus, if we start with the basic rule that fire service organizations, as an extension of the government, cannot be sued and then add the exceptions which have been carved out of the general rule by the courts, by the statutes, or by other rules that need to be applied to the situation, we usually can ascertain the position which a fire service organization is in following an incident. It should be noted that these myriad rules vary from state to state and from municipality to municipality and there may be a substantial number of variations to the rules of liability in any given jurisdiction. It should also be noted that our society is trending toward individual rights rather than governmental protections and thus new and novel theories for recovery for an injured individual may be also available.

Review of the cases in various jurisdictions which have considered the question of municipal liability for fire service organizations have revealed several general reasons leading to the conclusion that fire service organizations are either liable under a theory of tort liability or either immune from tort liability actions or inactions. In previous chapters, the question with regard to fire service liability under a tort theory revolved around whether the fire service organization had a duty to the individual. Below are some of the decisions which may provide guidance in this area.

1. Where a municipality is not required by statute to establish and maintain a fire service organization, there is generally no statutory obligation, and thus no duty, to protect any person or property.

2. Where a statute has been established by a legislature requiring a municipality to establish and maintain a fire service organization, a duty may exist to protect the public and/or an individual or individual's person or property.

3. Where contractual obligation to provide fire services is established, a duty is normally created through this contractual obligation. Absent a contractual obligation to provide these services, usually no duty to perform has been created. Fire service organizations, as an extension of a governmental entity, should be free to exercise their discretion and choice of tactics as long as appropriate, without worry of possible allegations of negligence in their decision making. As long as the fire service organization is within the scope of its authority and using professional judgment in its activities, states retain qualified immunity for these "governmental activities." In most jurisdictions, the doctrine of sovereign immunity is based upon the characteristics of the activity (i.e., governmental v. proprietary, congressional v. administerial). Most states have developed statutory protections for specified government activities. Statutory immunities are normally narrowly construed and will provide immunity only within the activities specified within the statute.

FIRE OFFICER'S LIABILITY

In the area of fire officer's liability, a definite distinction must be made between criminal liability and civil liability. In virtually all situations, the fire officer is responsible for his/her own criminal acts. In the area of civil liability, the lines are often blurred dependant upon the facts of the situation.

As a general rule for civil liability, fire service officers are usually provided the same immunity as the fire service organization. Under the concept of *Respondeat Superior*, so long as the fire service officer is functioning within the scope of his or her employment, the fire service organization usually is the named responsible party. However, when a fire service officer is outside the scope of employment or purposefully or willfully endangers another through their actions or omissions, personal civil and potentially criminal liability may result. Under many federal and state statutes, both civil and criminal liability may result where the officer willfully disregards a duty as proscribed by law (example: OSHA, EPA, and other regulations).

Additionally, as we are aware, civil actions are seeking monetary damages and the "deep pocket" in most civil actions is the fire service organization or municipality rather than the individual fire officer. However, when a fire officer is beyond the scope of duty and injury or harm occurs, potential personal liability may attach.

In a substantial number of fire service organizations, the fire service, through charter provisions or by-laws or by separate contract, indemnify their officers for errors and omissions made while performing within the scope of their employment. Indemnity, by definition, is "a collateral contract or assurance, by which one person engages to secure another against anticipated loss or to prevent him from being damnified by the legal consequences of an act or forbearance on the part of one of the parties or of some third party. Terms pertaining to liability for loss shifted from one person held legally responsible to another person."[1] In a fire service situation, indemnification means that the fire service organization will bear the costs of legal fees, any civil penalties, and damages for the officer if he/she is named personally in the action which is basically against the fire service organization or municipality.

The greatest potential for civil liability falls for fire service officers within the category of tort liability (i.e., negligence). With the ever expanding trends to provide exceptions to the immunity doctrine and our increasingly litigious society, fire service officers may increasingly become a target for civil actions just as officers and managers have in the private sector. Fire service officers should take the necessary precautions to protect their organization and themselves through appropriate planning, preparation, and decision-making.

---

[1]*Black's Law Dictionary,* West Publishing, 1983.

FIRE DEPARTMENT'S LIABILITY

The question of whether or not a fire department is liable tends to fall within the following categories:

1. Was the activity in which the harm occurred a proprietary function in which no judgment was required?
2. Was the activity in which the harm occurred a governmental function?
3. Was the activity in which the harm occurred within the category of judicial, quasi judicial, legislative, or quasi legislative function?
4. Was the activity in which the harm occurred outside of any/all of the above functions?

These categorizations are necessary in evaluating potential civil liability because of the potential protections afforded through immunity. However, as noted above, the concept of immunity is fast deteriorating due to the number of exceptions being provided by the courts.

The second category of questions which should be asked in evaluating potential civil liability surround the facts of the situation which resulted in the harm:

1. Was a duty created by statute or other law which required to fire service organizations to respond?
   a. Was a special duty created through actions?
   b. Was that duty breached?
2. What was the nature of the duty violated (proprietary—governmental—ministerial)?
3. Was the negligent individual a fire fighter, an agent, or otherwise employed by the fire service organization?
4. Was the fire fighter's act beyond the scope of his/her responsibility or authority?

5. If the fire fighter was acting outside the scope of authority, did the fire service organization or municipality subsequently ratify the act?
6. Was the fire service organization, its fire fighters, agents, or employees negligent?
7. Was the injured employee another fire fighter or employee of the fire service organization or municipality? (Do Workers Compensation laws apply?)
8. Did the injured individual assume the risk or extend the harm?

As with individual liability, the major area of potential liability for fire service organizations is under the theory of tort liability (i.e., negligence). Most fire service organizations have a duty created by statute, law, or the creation of a special relationship and when that duty is breached and harm occurs, the potential of civil liability is present. Activities such as fire suppression, tactics and strategies, and 911 system failures are only a few of those performed by fire service organizations, which are highly susceptible to potential liability when the activity malfunctions and harm occurs.

IMMUNITY

In most states, statutory immunity of varying types has been enacted to protect fire service organizations from civil liability. However, through performing activities outside of the statutory protection and exceptions which have been provided by the courts, this level of protection against civil liability is limited at best.

A review of the substantial number of cases where immunity was at issue identified the activities of operation of a fire service vehicle, fire suppression activities, hydrant and water supply, fire station and fire equipment, and torts by fire fighters to be the major areas where various courts analyzed this protection.

For fire service vehicles, the first factor generally considered was whether the injury was caused by the negligence of the operator of the fire service vehicle. Under the doctrine of *Respondeat Superior*, fire service organizations are generally responsible for the negligent acts or omissions of their employees operating fire service vehicles. However, many states have provided specific statutory immunity in special circumstances.

Some states have required the operator of the vehicle to be an employee of the fire service organization and be driving an authorized emergency vehicle in order to be afforded protection.[2] Other states provide requirements that the vehicle be responding to an emergency call or fire alarm and some states require the siren to be sounding and warning lights displayed. Given the expansive nature of the decisions with regards to fire service vehicles, prudent fire service organizations should become knowledgeable about the statutory protections provided, if any, and the case law in the jurisdiction regarding the operation of vehicles.

Of importance in mutual aid situations is the fact that the protections provided by statute do not normally go beyond the boundaries of the jurisdiction. Mutual aid agreements can be informal agreements or written agreements to provide assistance in emergency situations. The fire service organization must have the authority to enter into such agreements. Often mutual aid agreements are not considered binding unless the fire service organization negotiates the terms and conditions of the agreement. Of importance in any mutual aid agreement should be the protections provided under statutory immunity when the fire service organization is outside of the jurisdiction and indemnification and hold-harmless agreements in the event of personal injury or property loss.

In the areas of dispatch, extinguishment, and suppression activities, the general consensus of the courts is that the fire service organization is immune from liability.[3] However, where the fire service

---

[2]See, *Indiana Code* § 9-4-1(d).

[3]See, *Helman v. County of Warren*, 111 A.D.2d 560, 67 N.Y.S. 799 (1986).

organization is negligent to exercise reasonable care, liability can be found. Cases where fire service organization have been held liable include failure to respond to several 911 calls,[4] failure to respond to a fire call, and failure to inspect valves.

In the area of extinguishment of fires, the major factors usually considered are whether the failure to extinguish the fire is due to a failure of equipment, failure of suppression activities, or a complete failure to respond. In most jurisdictions, the failure of equipment and failure to suppress are usually covered under immunity where the failure to respond may provide risks of liability. Failure to provide sufficient water, failure to inspect fire hydrants and valves, and fire stations (as attractive nuisances) are also areas where risks of potential liability may be found where immunity protection may or may not be available.

Fire stations and fire apparatus are often categorized as "attractive nuisances" under the Attractive Nuisance Doctrine.[5] This doctrine, designed to protect small children, specified that a fire service organization, as the possessor of the property, may be liable for physical harm to children trespassing thereon caused by an artificial condition (i.e., bells, whistles, red paint) if:

1. The place where the condition exists is one upon which the fire organization knows or has reason to know that children are likely to trespass.
2. The condition is one which the fire service organization knows or has reason to know and which the fire service organization realizes or should realize involves an unreasonable risk of death or serious bodily harm to such children (for example, backing a vehicle from the station).
3. The children, because of their youth, do not discover the condition or realize the risk involved in intermeddling with it or in coming within the area made dangerous by it; and

---

[4]*Trezzi v. City of Detroit,*
[5]Restatement (Second) of Torts §339.

4. The fire service organization fails to exercise reasonable care to eliminate the danger or otherwise protect the children.[6]

SPECIAL DUTY DOCTRINE

Though fire service organizations are generally immune from liability for truly discretionary undertakings or determinations, that exemption may be nullified where a special relationship has developed between the fire service organization and the injured individual (Chapter 3). Typically, this occurs when a fire service organization undertakes a discretionary activity (thus producing a detrimental reliance by the individual for the fire service organization to continue or completely perform an activity). The fire service organization subsequently ceases performance or does not perform without notifying the individual, the theory of detrimental reliance upon the activity is thus established.

There are several factors courts consider in determining whether the creation of a new duty relationship has been established in which an individual may rely on their detriment to a fire service organization. The courts in several jurisdictions have found that a special relationship exists and thus a special duty created by the fire service organization in the following circumstances:

1. For an individual to receive the benefits of the fire service organization's actions are reciprocally identified and have a direct contact with the fire service organization. The individual must be readily distinguished from the public as a whole.
2. The fire service organization was in a position of superior knowledge or authority, which could reasonably induce reliance by the individual on the statements of the fire service organization or its agents.
3. If the cities fire safety code requirement required the fire service organization to perform specific acts upon the finding

---

[6]See, for example, *Loney v. McPhillips*, 521 P.2d 340 (Or. 1974).

upon a hazard, failure to perform these acts were considered gross negligence.

4. Fire service organizations communicating the assumption of corrective actions obligated the fire service organization to the individual. This communication can either be by words or by actions to establish a special duty to the fire service organization to protect the individual. The fire service organization will be considered to owe a particular individual a special duty of care when its officers or agents, in a position of authority to act or not to act, have or should have had knowledge of a condition that violates safety standards prescribed by statute or regulation and that presents serious risk of harm to a specified individual or property. When a fire service organization can reasonably foresee this danger, the fire service organization has a duty to exercise reasonable care for that specific individuals benefit.

In our world today, fire service organizations have the potential for tort liability, and immunity of any type is the exception rather than the rule. In order for fire service organizations to evaluate the potential risk from tort liability in any given situation, the following questions have been provided to perform a simple analysis of the risks in any given situation:

1. Has the duty been created by statute or duty actions or inactions of the fire service organization to respond to an individual's need?

2. Was the duty violated during a proprietary rather than a governmental activity, or with an administerial rather than a discretionary function?

3. Was the individual performing the negligent act or omission an employee or agent of the fire service organization?

4. Was the act *ultra vires* (beyond the individuals scope of authority or responsibility as provided by the fire service organization)?

5. Was the fire fighters act subsequently ratified by the acts of the fire service organization or municipality?
6. Does the fire service organization or municipality have a duty to protect the particular individual that incurred the injury or property damage reasonably and detrimentally relied upon the affirmative representations of the fire service organizations, the fire fighters, or the agents of the fire service organization?
7. Did the fire service organization, fire fighters, or its agents perform the duty required by statute?
8. Was the fire service organization guilty of negligence and was the employee free from contributory negligence?
9. Was the assumption of the risk doctrine applicable?
10. Is the fellow servant rule applicable?
11. Do the provisions of the individual state's workers compensation laws apply to the situation?
12. Does the individual state's statutory immunity apply to this situation?

The law of torts can be categorized by either (1) the nature of the defendant's conduct (i.e., intentional torts, negligence, product liability) or (2) the nature of the harm resulting to the injured individual, the damage to the property, or the harm to intangible personal interest.

Intentional Harm to Persons

The torts involving intentional harm to persons are assault, battery, false imprisonment, and intentional infliction of emotional distress.

Assault

The components of an assault include the intent to commit a harmful or offensive touching or creating apprehension of the same; to the apprehension of offensive touching or harm in which the subjective mental state is in question. To establish the tort of assault, the plaintiff must prove that he was placed in apprehension of imminent harm or offensive touching. This is a subjective standard. An act which is intended to put another in apprehension of immediate bodily harm and succeeds in doing so, may be considered assault. Assaults can result if the plaintiff is unaware of the threat and words alone cannot make an individual liable for an assault unless, together with other acts or circumstances, they put the plaintiff in reasonable apprehension of reasonable harm or offensive contact. It should be noted that a touching is not required for an assault.

Battery

The interest addressed in a battery action is the right of a person to be freed from intentional or offensive bodily contact. The basic components of a battery include (1) the intent to commit harmful or offensive touching or creating an apprehension of the same, (2) an actual touching of the individual, and (3) an absence of consent by the individual who was touched. In a battery action, the intent may be transferred from one individual to another and the direct application of force may be used. The touching must be of a harmful or offensive nature. The touching, which is not permitted by the norms of modern custom even though not harmful in nature, will suffice in most circumstances[7] (i.e., removal by the defendant of the plaintiff's false teeth, by force). The contact is offensive if it offends a persons reasonable sense of personal dignity (i.e., spitting upon or slapping a person). An individual being touched must be aware of the touching (i.e., if the individual is asleep or unconscious) and must not consent to the touching. For a touching to actually be battery, the plaintiff must establish that he/she did not consent to the touching. A touching obtained by duress is ineffective. Only consent obtained by fraud, misrepresentation or mistake

---

[7] See, *Jones v. Fisher*, 42 Wisc. 2d. 209, 166 N.W. 2d. 175 (1969)

is ineffective. In these situations, the traditional rules that a doctor may extend an operation only when emergency exists in that a failure to extend the operation would endanger the patient's life or health, or whether a later operation might unduly endanger the life or the health of a patient or is impractical to obtain the consent of the patient's family.[8] The states have adopted a more liberal view in light of the conditions in which the operations are now performed. In *Kennedy v. Parrott*,[9] the court held that in the absence of proof to the contrary, consent to an operation will be construed as general in nature, and the surgeon may extend the operation to remedy any abnormality or disease conditions in the area of the original incision wherever he, in the exercise of his sound professional judgment, determines the correct surgical procedures dictate and require that an extension of the operation beyond the originally contemplated period. The test applied by most courts is whether a reasonable person in custodial care of the patient would grant consent to the extension if he were able to choose.

The area of malpractice involving the extension of an operation without the consent of the patient (where an action sounds like a battery because the absence of consent makes the touching unpermitted) should be compared with the medical malpractice where a patient claims that he/she consented to a medical treatment by the doctors withholding of material information concerning the risk of such treatment. In the United States today, lack of consent cases are treated as negligent cases rather than battery.

False Imprisonment

The interest protected by false imprisonment is the individual's freedom from confinement or from restraint of movement. The elements of the false imprisonment action include (1) intent, (2) confinement, (3) consciousness of confinement, and (4) absence of

---

[8] *Tabor v. Scobee*, 254 S.W. 2d. 474 (Ky. 1953)).

[9] 243 N.C. 355, 90 S.E. 2d. 754 (1956).

consent. False imprisonment is an intentional tort (i.e., there is no such tort as a negligent false imprisonment tort). As with the action for battery, a distinction must be made between intent and motive. Absent privilege, if a defendant intentionally confines a plaintiff without his/her consent, the plaintiff may recover even if the person acted with good motive. In the area of confinement, the jury find that the defendant imposed a restraint on his movement. An action will not lie in false imprisonment where a defendant merely impedes the individuals progress in one direction or confines him within an area where there is a reasonable exit. Insufficient false imprisonment may result by taking a person into custody under a assertive lawful authority which was in fact unlawful. No force or threat of force is necessary. The individual must be conscious in order to knowingly be confined. As with all of the intentional torts, the false imprisonment action also lacks consent.

Intentional Infliction of Emotional Distress

"One who, by extreme or outrageous conduct, intentionally or recklessly causes severe emotional distress to another, is subject to liability for such emotional distress and if bodily harm results from it, for such bodily harm." Restatement, Second Tort Section 46. The elements for intentional infliction of emotional distress include intent, the impact of the distress must be severe, and a special relationship must exist between the parties. In most states, the intentional infliction of emotional distress may be based not only the subjective intent or substantial certainty of the result but also on reckless disregard of probable consequences of the defendants behavior. Additionally, in most states, the emotional distress must manifest itself in a severe form. Causing embarrassment or humiliation are normally inadequate for emotional distress. A special relationship is required between the parties to have intentional infliction of emotional distress. Courts allow recovery from common carriers, innkeepers, and public utilities for emotional distress to patrons resulting from gross insults of a highly offensive nature.[10] The liability for insulting language has been extended to owners of a place of business open to the public.[11] Liabilities for emotional distress can be inflicted on bystanders to the situation. Whether or not the defendant directs extreme or offensive conduct at a third person, he may be responsible for liability to the third person if he intentionally or recklessly causes severe emotional distress to a member of the immediate family who was present at the time whether or not such distress results in bodily harm to the third person who is present at the time if such distress results in bodily harm.[12]

The defenses to intentional torts are varied. These defenses are usually available to actions of intentional torts to individuals who are privileged or who have immunity. A privilege is a justification to a defendant's conduct which actually negates tortious quality. An immunity goes not to the nature of the conduct but only to whether or not the defendant may be sued for such conduct. In the area of privilege, the defenses include self defense, defense of others, defense of property, and arrest without a warrant. In the area of immunity, the defenses include interspousal immunity, parent and child immunity, charitable immunity and governmental immunity. As a defense, a person who is threatened with bodily harm may meet his aggressor's force to repel the attack. This means that an individual may use whatever force necessary to repel an attack from an aggressor. A person's right to use force in self defense is not limited to the situation where he is limited to injury or death.[13] However, unless the defendant was attacked with force apparently sufficient to warrant death or serious bodily harm, he may not use self defense force calculated to cause death or serious bodily harm (Restatement, Second Tort Section 65). In some states, the individual has the obligation to retreat when being attacked with deadly

---

[10]See, *Lipman v. Atlantic Coastline Railroad Company,* 108 Sup. Ct. 151, 93 S.E. 2d. 714 (1917)); Restatement (Second) of Tort § 48.

[11] See, *Slocum v. Food Fair Stores, Florida,* 100 S. 2d. 396 (FL. 1958)).

[12] Restatement (Second) of Torts § 46.

[13]*Boston v. Muncy,* 204 Ok. 603, 233 P.2d. 300, 25 A.L.R. 1208 (1951).

force. With defense of a third person, a person is privileged to use reasonable force in defense of this individual. It does not generally require, as a condition of this privilege, that a person being defended be a member of the actor's household or one who is under a legal or social duty to protect. Where the actor or the individual makes a mistake as to the need to defend the third person, states are split on the issue of liability. The majority view is that the actor takes the risk that the person he is defending would not be privileged in defending himself in like manner. The minority view is that the standard of liability for the actor who mistakenly goes to the defense of a third person should be the same as when he makes a mistake in believing that self defense is necessary. Reasonable mistake would preclude liability. Thus, the amount of force to be used by the individual assisting the third person would be to the individual being attacked. For defense of property, force is not permitted. There is no privilege to use any force calculated to cause death or serious bodily harm to repel a threat to land or chattel unless there is also a threat to the person or others safety.[14] Force, in the recapture of chattel (personal property), is permitted when there is a wrongful taking and there is fresh pursuit. A common situation involving the use of force to recapture property wrongfully is a shoplifting situation. The law usually recognizes the privilege of merchants to detain a person when the merchant has reasonable grounds that the person is stealing or attempting to steal his property. The detention, however, must be a reasonable period of time and must be conducted in a reasonable manner. The purpose of this privilege is to recover the goods, and after the goods are recovered, the defendant continues to detain the individual to obtain a signed confession. The action for false imprisonment may follow.[15] In most states, merchant privilege is normally covered by statute.

---

[14] *Katko v. Briney*, 183 N.W. 2d. 657 (Iowa 1971).

[15] See, *Totel v. May Department Store Company*, 348 Mo. 696, 155 S.W. 2d. 74 (1941)).

SELECTED CASES

(This case has edited for the purposes of this text.)

Holger A. **Holmberg**
v.
Douglas **Brent** No. 92-300
Supreme Court of Vermont
Nov. 19, 1993

Before Allen, C.J., and Gibson, Dooley, Morse and Johnson, JJ.

In March 1987, the plaintiff, Holger Holmberg, was injured while employed as a fire fighter by the Village of Bellows Falls Fire Department, when, in response to an emergency call, he slid down a fire pole and landed on the cement floor below. Sometime before the incident, the defendant, Douglas Brent as fire chief, had ordered the removal of a pad surrounding the base of the pole. The plaintiff brought a negligence action, alleging that removing the pad had created an unreasonably dangerous condition and caused his injury. Plaintiff received workers compensation benefits, but sought damages from defendant under 21 V.S.A. s 624, which permits suits against third parties responsible for injury.

The defendant moved for summary judgment, alleging that he qualifies as a municipal officer under 24 V.S.A s 901(a), which would require the plaintiff to bring his action against the Village of Bellows Falls instead of the defendant. The defendant argued in the alternative that qualified official immunity shields him from any liability, because removal of the pad was a discretionary function of the office of fire chief.

The trial court entered judgment for the defendant without ruling on the qualified immunity issue. As a result, the plaintiff's only recourse would be to sue the village, his employer, directly for the fire chief's alleged negligence, but such an action is barred under the exclusive remedy provision of the workers compensation statutes. See 21 V.S.A. s 622. In effect, the judgment would deprive the plaintiff of a remedy under the workers compensation exclusivity exception.

The plaintiff contends that the trial court erred in ruling that no material fact was disputed and that the defendant should prevail under a proper interpretation of s 901(a). Summary judgment may be granted if no genuine issue of material fact exists and the moving party is entitled to judgment as a matter of law.

In relevant part, 24 V.S.A. s 901(a), entitled, "Actions by or against town officers" requires that "[w]here an action is given to any appointed or elected municipal officer...the action shall be brought in the name of the town in which the officer serves...If the action is given against such officer, it shall be brought against such town...."Despite the apparently limiting reference to towns, the legislature has declared that "the laws applicable to the inhabitants and officers of towns shall be applicable to the inhabitants and similar officers of all municipal corporations." The Village of Bellows Falls, an incorporated village qualifies as a municipal corporation because the term "municipality" includes incorporated villages. 1 V.S.A. s 126. Therefore, assuming the fire chief is a "municipal officer" 24 V.S.A. s 901 would apply to the village and require that the action be brought against the village, unless s 901 conflicts with a law specifically applicable to the village. See 1 V.S.A. s 139.

One such specific law is 24 V.S.A. s 1313, which provides that:

An incorporated village, by vote, may indemnify a duly appointed public or peace officer of the village against legal proceedings for injuries committed by him while in the lawful discharge of his official duties. If an action is commenced against such village, the trustees may defend such action at its expense.

The distinguishing characteristic of a "public officer" is that the officer carries out a sovereign function. Black's Law Dictionary 1230 (6th ed. 1990). For purposes of this argument we adopt the defendant's assertion that the fire chief's duties "involve the continuous exercise of sovereign power in the public interest," which fits the definition of a "public officer." Therefore, as an appointed official, Bellows Falls Charter s 18, the fire chief is a "duly appointed public officer" covered by the provisions of 24 V.S.A. s 1313.

Since the defendant is a "public officer" subject to s 1313, he may be sued in his individual capacity as a fire chief. The defendant is not entitled to summary judgment under 24 V.S.A. s 901(a).

We express no opinion on the defendant's claim that he is entitled to qualified official immunity, as the trial court did not consider the issue in its ruling on the summary judgment motion.

Reversed and remanded.

(This case has been edited for the purposes of this text.)

James Quentin **Taylor**
v.
Terry Kenneth **Ashburn**
Court of Appeals of North Carolina
Nov. 16, 1993

On September 25, 1989, plaintiff (Taylor) was driving his automobile on a road in North Carolina. Around the same time, defendant (Ashburn), a fire engineer, was operating a fire truck owned by the City of Winston-Salem (the City) and was responding to a call at a high rise housing complex. The fire truck had its emergency equipment—siren, flashing lights, and horn—in full operation. At an intersection, plaintiff's automobile and the fire truck driven by the defendant collided.

On January 21, 1992, plaintiff filed a complaint against defendant alleging that the accident was the result of defendant's negligent operation of the fire truck and caused plaintiff "substantial bodily injury, property loss, loss of income and other incidental damages." In his complaint, plaintiff did not specify whether he was suing the defendant in his individual capacity or in his capacity as a fire engineer for the City. Plaintiff alleged in his complaint that, " ... [d]efendant was operating a ... fire truck owned by the City of Winston-Salem and was operating said vehicle with the permission of the City ... in connection with his employment as a fireman and was in the course and scope of his employment and agency."

On April 9, 1992, defendant filed an amended answer pleading the affirmative defenses of governmental immunity for any claims resulting in damages up to and including $1,000,000, and of immunity from liability for acts committed in the course and scope of defendant's capacity as a public officer. On October 5, 1992, based on his affirmative defenses, filed a motion for summary judgment which was denied by the trial court.

The issue on appeal is whether the plaintiff's complaint, which alleges that defendant was operating a fire truck in the course and scope of his employment as a fireman for the City when the accident occurred, constitutes suing the defendant in his official capacity so that he shares in the City's governmental immunity.

Under the doctrine of governmental immunity, a municipality is not liable for the torts of its officers and employees if the torts are committed while they are performing a governmental function, which includes the organization and operation of a fire department. Governmental immunity protects the governmental entity and its officers or employees sued in their "official capacity." Although a plaintiff generally designates in the caption of his or her complaint in what capacity a defendant is being sued, this caption is not determinative on whether or not a defendant is actually being sued in his or her individual or official capacity. The court must inspect the complaint as a whole to determine the true nature of the claim. If the plaintiff fails to advance any allegations in his or her complaint other than those relating to a defendant's official duties, the complaint does not state a claim against a defendant in his or her individual capacity, and instead, is treated as a claim against defendant in his official capacity.

In this case, plaintiff's complaint does not mention the words "individual capacity" or the words "official" or "official capacity." Plaintiff alleges that "[d]efendant was operating . . . a fire truck owned by the City . . . with the permission of the City . . . in connection with his employment . . . and was in the course and scope of his employment and agency." The allegations in plaintiff's complaint concern only defendant's actions while performing his official duties as a fire engineer of driving a fire truck for the City and responding to an emergency call. After review of this language and the complaint as a whole, we hold that plaintiff has asserted a negligence claim against defendant in his official capacity alone. Therefore, defendant shares in the City's governmental immunity. Because defendant has met his burden of showing that plaintiff cannot surmount defendant's affirmative defense of governmental immunity which bars plaintiff's claim, the trial court erred in denying defendant's motion for summary judgment. We remand for entry of summary judgment for defendant.

Reversed and remanded.

Martin and **John**, J.J., concur.

5

Criminal Liability for Fire Fighters

Litigant: a person about to give up his skin for the hope of retaining his bone.

Ambrose Bierce

In America, an acquittal doesn't mean you're innocent, it means you beat the rap. My clients lose even when they win.

F. Lee Bailey

OVERVIEW OF POTENTIAL LIABILITY

Fire fighters are responsible for their own criminal acts. If the act performed by the fire fighter is in violation of a criminal law, the individual fire fighter will be the responsible party. If the fire service organization is responsible for the criminal act, then the officers and board may be responsible for the crime. Fire fighters are *not* immune from the committing of criminal acts and, in fact, because of the duties of the job, fire fighters are often placed in positions of trust where criminal activity can easily be performed (such as theft during a fire, arson, and related crimes) to those so inclined. In essence, fire fighters are responsible for compliance with all laws and are subject to noncompliance just as any other citizen. For example, if a fire fighter intentionally kills a person at a fire scene or at any other location, the individual fire fighter may be liable for murder under the state criminal code.

Criminal laws can be either federal or state. In most circumstances, the location of the criminal act or the type of act is the deciding factor as to whether the crime would be prosecuted by state or federal authorities. For example, if a murder was committed on a city street, the murder would usually be prosecuted by the state authorities. If the murder was committed on federal property, the murder charge would be prosecuted by federal authorities. Certain crimes, such as crimes involving interstate actions (like mail fraud) usually are prosecuted by federal authorities.[1] Each particular crime has elements which must be proven by the federal or state prosecutor. The burden of proof is on the federal or state prosecutor to prove each element. Trial by jury is usually permitted in all criminal cases and the penalty for the crime is usually prescribed by law.[2]

A relatively new phenomena which could substantially impact the scope of potential criminal liability for fire service organizations is the use of state criminal laws by state and local prosecutors for injuries and fatalities which occur on the job. This new utilization of the standard state criminal laws in a workplace setting normally governed by OSHA or state plan programs is controversial but appears to be a viable method through which states can penalize officers in situations involving fatalities or serious injury. At this point in time, this new utilization of state criminal laws does not appear to be preempted by the OSH Act. It should be clarified that the use of criminal sanctions for workplace fatalities and injuries is not a new area of concern. Under the OSH Act, criminal sanctions have been available since the inception in 1970. In Europe, use of criminal sanctions for workplace fatalities are frequently used and, in the U.S. as far back as 1911, criminal sanctions were used in the well-known Triangle Shirt fire in New York which killed over 100 young women. In this case, the owners of the Triangle Shirt Company were indicted on criminal manslaughter charges although subsequently acquitted of these charges. Fire officers should, however, take note that the sources of the potential criminal liability (i.e., state criminal codes in addition to the OSH Act) and the enforcement frequency (i.e., increase use of criminal charges under the OSH Act and state criminal codes) is a recent trend.

---

[1] On the federal level, the United States Justice Department is usually the prosecuting authority.

[2] See, for example, the Federal Sentencing Guidelines, 18 U.S.C. § 3577.

The case which propelled the issue of whether OSHA had jurisdiction over workplace injuries and fatalities (and thus preempts state prosecution under state criminal statutes for workplace deaths) was the murder charges brought about by the Cook County prosecutor against the president, the plant manager, and the plant foreman of Film Recovery Systems, Inc. for the 1983 work-related death of an employee.[3] In this case, Stephan Golab, a 59-year-old immigrant employee from Poland, died as a direct result of his work at this Elk Grove, Illinois manufacturer in which he stirred tanks of sodium cyanide used in the recovery of silver from photographic films. In February 1983, Mr. Golab "walked into the plant's lunchroom, started violently shaking, collapsed and died from inhaling the plant's cyanide fumes."[4] Following his death, both OSHA and the Cook County prosecutor's office investigated the accident.

OSHA found 20 violations and fined the corporation $4850.00. This monetary penalty was later reduced by half.[5] The Cook County prosecutor's office, on the other hand, took a different view and filed charges of first degree murder and 21 counts of reckless conduct against the corporate officers/management personnel and involuntary manslaughter charges against the corporation itself. [6] Under the Illinois law, murder charges can be brought when someone "knowingly creates a strong probability of death or great bodily harm" even if there is no specific intent to kill.[7] The prosecutor's office initially brought charges against five officers and managers of the corporation but the defendant successfully fought extradition from another state.[8] The prosecutor's intent was to

---

[3]*People v. O'Neil, et. al. (Film Recovery Systems)*, Nos. 83 C 11091 & 84 C 5064 (Cir. Ct. of Cook County, Ill. June 14, 1985), rev'd, 194 Ill. App.3d 79, 550 N.E.2d 1090 (1990).

[4] Gibson, *A Worker's Death Spurs Murder Trial*, The National *Law Journal*, January 21, 1985 at page 10.

[5] Note: *Getting Away With Murder—Federal OSHA Preemption Of State Criminal Prosecutions for Industrial Accidents*, 101 Harv. L.J. 220 (1987) (The violations ranged from failure to instruct employees about the hazards of cyanide and to provide first-aid kits with antidotes for cyanide poisoning to failure to keep the floors clean and dry. OSHA did not cite Film Recovery Systems, Inc. for exceeding the permissible exposure limit for cyanide. See, Citations and Notification of Penalty issued by O.S.H.A. to Film Recovery Systems, Inc., Report #176 (Mar. 11, 1983) and Amendment (Mar. 30, 1983).

"criminally pierce the corporate veil" to place liability not only upon the corporation but on the responsible individuals.[9] During the course of the trial, the Cook County prosecutor presented extensive and overwhelming testimony that the company officials knew of unsafe conditions, knowingly neglected the unsafe conditions, and attempted to conceal the dangers from employees. Witnesses testified to the following:

- Company officials exclusively hired foreign workers who were not likely to complain to inspectors about working conditions and would perform work in the more dangerous areas of the plant.
- Officials instructed support staff never to use the word cyanide around the workers.
- According to the testimony of two workers, Mr. Golab complained to plant officials and requested to be moved to an area where the fumes were not as strong shortly before his death. These pleas were ignored.

---

[6] Id.; Also see, *Brief for Appellee* at 12-20; [The state presented evidence that the deceased employee and many of the other Film Recovery system employees who worked around unventilated tanks often experienced headaches, nausea, and burning eyes and skin from cyanide fumes. The plant management never informed the employees that they were working with a deadly toxin and provided them with virtually no safety equipment. Most of the workers were illegal aliens, primarily from Mexico and Poland, and thus could not read the warnings and were vulnerable to losing their jobs if they complained] and Gibson, *A Worker's Death Spurs Murder Trial,* The National Law Journal, January 21, 1985 at page 10; [Ms. Erpito (secretary at Film Recovery Systems, Inc.) said she was instructed by her boss never to use the word cyanide around workers. She testified that when going into the plant, "your eyes started to burn and you got a headache"].

[7] Moberg, David, et. al., *Employers Who Create Hazardous Workplaces Could Face More Than Just Regulatory Fines, They Could Be Charged With Murder,* 14 Student Lawyer, Feb. 1986 at page 36. (The legislative intent may have had cases like arsonist torching in mind but the principle is easily extended to cases such as Film Recovery Systems, Inc.).

[8] Gibson, *A Worker's Death Spurs Murder Trial,* The National Law Journal, January 21, 1985 at page 10. (The original indictment before the grand jury in October 1983 was against Steven J. O'Neill, the former president; Michael McKay, an officer; Gerald Pett, vice President; Charles Kirschbaum, plant manager; and Daniel Rodriguez, a plant foremen. Mr. McKay successfully fought extradition from Utah in February 1984).

[9] Moberg, David, et. al., *Employers Who Create Hazardous Workplaces Could Face More Than Just Regulatory Fines, They Could Be Charges With Murder,* supra at page 36.

- Testimony of numerous employees stated recurrent nausea, headaches, and other illnesses were common. Employees were not issued adequate safety equipment or warned of the potential dangers in the workplace.
- An industrial saleswoman reported trying unsuccessfully to sell safety equipment to the owners.
- That the supervisors instructed employees to paint over the skull-and-crossbones on the steel containers of cyanide-tainted sludge and to hide the containers from inspectors after the employee's death.
- Workers were never told they were working with cyanide and were never told of the hazardous nature of this substance.
- Employees were grossly overexposed on a daily basis and, after the incident, the company installed emission-control devices which dropped cyanide emissions twenty-fold.[10]

Given the extreme circumstances in this case, the prosecution was able to obtain conviction of three corporate officials for murder and 14 counts of reckless conduct, and each was sentenced to 25 years in prison. The corporation was convicted of manslaughter and reckless conduct and fined $24,000.[11] The court rejected outright the company's defense of preemption of the state prosecution by the Federal Occupational Safety and Health Act. This case opened a new era in criminal liability and, as the prosecutor appropriately stated: "These verdicts mean that employers who knowingly expose their workers to dangerous conditions leading to injury or even death can be held criminally responsible for the results of their actions." "Today's [criminal] verdict should send a message to employers and employees alike that the criminal justice system can and will step in to protect the rights of every worker to

---

[10]*People v. O'Neil, et. al. (Film Recovery Systems)*, Nos. 83 C 11091 & 84 C 5064 (Cir. Ct. of Cook County, Ill. June 14, 1985), **rev'd**, 194 Ill. App.3d 79, 550 N.E.2d 1090 (1990).

[11] Sand, Robert, *Murder Convictions For Employee Deaths; General Standards Verses Specific Standards*, 11 Employee Rel. J. 526 (1985-86).

a safe environment and to be informed of any hazard that might exist in the work place."[12]

As expected, the decision in *People v. O' Neil* opened a new era in work place health and safety. [13] The case marked the first time in which a corporate officer had been convicted of murder in a work place death. As expected, once the door was opened, prosecutors across the country began to initiate similar action, such as in the Imperial Foods fire in which the plant owner plead guilty to manslaughter and received a 20 year prison sentence,[14] and institute programs, such as the Los Angeles County District Attorney's "roll out" program,[15] to address work place accidents.[16]

Fire service organizations and their officers are responsible for the safety and health of the fire fighters under their direction. Where a fire service organization willfully places a fire fighter or others in harms way and an injury or fatality occurs, the potential of criminal liability has now become a reality.

OTHER CRIMINAL LIABILITY

Fire service personnel must be aware that they must comply with the law just as every other citizen in the United States. However, every year criminal charges are brought against fire fighters for

---

[12] Ranii, David, *Verdict May Spur Industrial Probes,* Nat. L. J., July 1, 1985 at p. 3.

[13] *People v. O'Neil, et. al. (Film Recovery Systems),* Nos. 83 C 11091 & 84 C 5064 (Cir. Ct. of Cook County, Ill. June 14, 1985), **rev'd**, 194 Ill. App.3d 79, 550 N.E.2d 1090 (1990).

[14] Jefferson, *Dying for Work,* ABA Journal, Jan. 1993 at p. 48.

[15] Id.

[16] Middleton, Martha, Get Tough On Safety, Nat L. J., April 21, 1986 at p. 1. [Note: Austin, Texas, prosecutors filed criminal charges in two cases in which workers were killed in trench cave-ins; Los Angeles district attorney Ira K. Reiner ordered a new "roll out" program in which an attorney and an investigator would be sent to the scene of every industrial workplace death; Milwaukee County District Attorney E. Michael McCann ordered investigators to begin checking every workplace death for possible criminal violations; U.S. Secretary of Labor William E. Brock referred the case of Union Carbide Corp. (pesticide violations in Institute, W. Va. facility) to the Justice Department. **Also see**, Sand, Robert, *Murder Convictions For Employee Deaths: General Standards Verses Specific Standards,* 11 Employee Rel. J. 526 (1985-86) [Cook County prosecutors initiated another action immediately after this case against five executives of a subsidiary of North American Philips (Chicago Wire)].

myriad charges ranging from auto theft to murder. Two areas of particular importance to fire service organizations and fire fighters are the areas of arson and domestic violence.

Given the specialized expertise of most fire fighters in the area of fire science, arson is often the weapon of choice by individuals within the fire service. Several alarming cases have evolved where fires have been intentionally set by individuals with fire service experience to cover other crimes and even to provide additional fire runs for the department. Prudent fire service organizations should not automatically exempt members of their organization when investigating an arson fire.

The second area is the violence which can erupt when a fire fighter is placed under an intense amount of stress because of the duties of the job. Incidents of homicide/suicide, domestic violence, and other criminal acts may be the result of pressures being placed upon the fire fighter on the job or away from the workplace. Utilization of debriefing sessions after an incident and employee assistance programs often help the member of the fire service organization cope with these stressors prior to a potentially violent release.

SELECTED CASES

(This case has been edited for purposes of this text.)

The **People** of the State of Illinois, Plaintiff-Appellee,

v.

Steven **O'Neil**, Film Recovery Systems, Inc., Metallic Marketing Systems, Inc., Charles Kirschbaum, and Daniel Rodriguez, Defendants-Appellants.
Appellate Court of Illinois
First District, Fifth Division
Jan. 19, 1990
194 Ill App. 3d 79, 550 N.E. 2d 1090 (1990)

Lorenz, J.

Individual defendants O'Neil, Kirschbaum, and Rodriguez, agents of Film Recovery Systems, Inc. were convicted of murder in the death of Stephan Golab, an employee of Film Recovery, from cyanide poisoning stemming from conditions in Film Recovery's plant. Corporate defendants Film Recovery and its sister corporation Metallic Marketing were convicted of involuntary manslaughter in the same death.

Film Recovery was in the business of extracting silver from used x-ray and photographic film for resale. The recovery process was performed by "chipping" the film and soaking the granulated pieces in large, open bubbling vats containing a solution of water and sodium cyanide. The cyanide solution caused silver contained in the films to be released.

On February 10, 1983, shortly after he began stirring the contents of a tank containing this cyanide solution, Stephan Golab became dizzy and faint. Mr. Golab left the production area to rest in the lunchroom. Workers present on that day testified that Mr. Golab's body trembled and he was foaming at the mouth. Golab eventually lost consciousness and was taken outside. Paramedics were unable to revive Mr. Golab and he was pronounced dead on arrival at the hospital.

After a toxicological report was received by the Cook County Medical Examiner, it was determined that Mr. Golab died from acute cyanide poisoning through the inhalation of cyanide fumes in the plant air.

Defendants were subsequently indicted by a Cook County grand jury. Defendants O'Neil, Kirschbaum, and Rodriguez were charged with murder. The indictment stated that as officers and high managerial agents of Film Recovery, the had knowingly created a strong probability of Golab's death. The indictment also stated that the individual defendants failed to disclose to Golab that he was working with substances containing cyanide and failed to advise him about, train him to anticipate, and provide adequate equipment to protect him from, attendant dangers involved. Film Recovery and Metallic Marketing were charged with involuntary manslaughter.

The indictment stated that through the reckless acts of their officers, directors, agents, and others all acting within the scope of their employment, the corporate entities had unintentionally killed Golab. Both corporate defendants and individual defendants were also charged with reckless conduct.

On June 14, 1985 the trial judge pronounced his judgment of defendants' guilt. The trial judge found that "the mind and mental state of a corporation is the mind and mental state of the directors, officers and high managerial personnel because they act on behalf of the corporation for both the benefit of the corporation and themselves." Further "If the corporation's officers, directors, and high managerial personnel act ... for their benefit and for the benefit of the corporation, the corporation must be held liable for what occurred in the workplace."

Individual defendants O'Neil, Kirschbaum and Rodriguez each received sentences of 25 years in prison for murder and 14 concurrent 364 day imprisonment terms for reckless conduct. O'Neil and Kirschbaum were also fined $10,000 each with respect to the murder convictions and $14,000 each with respect to the convictions for reckless conduct. Corporate defendants Film Recovery and Metallic Marketing were each fined $10,000 for the involuntary manslaughter convictions and $14,000 each for the reckless conduct convictions.

Defendants filed timely notices of appeal. One of defendants' principal contentions on appeal was that the federal Occupational Safety and Health Act of 1970 (29 U.S.C. § 651 *requisite*) preempted state criminal prosecutions against individual and corporate defendants for conditions in an industrial workplace. That identical issue was a subject of a then pending appeal before the Illinois Supreme Court in *People v. Chicago Magnet Wire*, 126 Ill. 2d 356, 128 Ill. Dec. 517, 534 N.E. 2d 962 (1989). Accordingly, we postponed disposition in this case until the Supreme Court decided *Chicago Magnet Wire*. The court in *Chicago Magnet Wire* concluded that such prosecutions *were not* preempted under the OSH

Act. Thus, we invited the parties to file new briefs and scheduled further oral argument on the remaining issues.

On appeal, the defendants urge that their convictions must be reversed and the cause remanded for retrial because the judgments rendered were inconsistent. Defendants also contend that the evidence presented at trial was insufficient to support the convictions.

Because the offenses of murder and reckless conduct require mutually exclusive mental states, and because we conclude the same evidence of the individual defendants' conduct is used to support both offenses and does not establish separately, each of the requisite mental states, we conclude that the convictions are legally inconsistent. Therefore, we now reverse those convictions as to both the individual and corporate defendants and remand the matter for retrial.

Reversed and Remanded.[17]

---

[17]Note: This case has been substantially edited for publication in this text. That portion of the court's reasoning with regards to the "requisite mental states" for murder and reckless conduct has been omitted.

6

Liabilities for the Safety of Fire Fighters

Well, I don't know as I want a lawyer to tell me what I cannot do. I
hire him to tell me how to do what I want to do.

J.P. Morgan

FIRE SERVICE ORGANIZATIONS' SAFETY RESPONSIBILITIES

Fire service organizations owe a duty, under the concept of Master
and Servant, to fire fighters and other employees to create and
maintain a safe and healthful work environment. This duty of care
extends not only to the physical locations but also to equipment,
inspection and repair, establishment of appropriate standard operating procedures (SOP), workplace rules and regulations, and
beyond. Given a fire service organizations varied workplace, this
duty to provide a safe working environment can extend to basically
anywhere in which a fire fighter may be assigned to work from the
top of a building to below the ground.[1] Fire service organizations
have an affirmative duty to create a safe and healthful work environment under the Occupational Safety and Health Act (discussed
below) and other state and local laws (such as the Public Liability
Act). Failure to comply with these laws can result in fines and/or

---

[1]See, *Barger v. Mayor and City Council of Baltimore*, 616 F.2d 730 (1980)(Duty owed is not limited
to fire apparatus and fire stations).

penalties, and where a fire fighter is injured or killed, the potential of criminal liability.

Fire service organizations should also be aware that they are being judged on the standards developed and published by the National Fire Protection Association (known as NFPA). Although these standards developed by NFPA are advisory in nature (such as NFPA 1500), after an accident has occurred, courts often use the applicable NFPA standards as the basis for establishing the standard of care (see Chapter VII).

Labor Organizations representing fire fighters also has an affirmative duty to create a safe and healthful work environment under various public employee relations and labor laws. Safety and health issues are usually considered a mandatory item in collective bargaining in the fact that safety and health of fire fighters often constitutes a term or condition of employment.

ASSUMPTION OF THE RISK DOCTRINE

The Assumption of the Risk Doctrine generally bars recovery by a fire fighter who has accepted an unreasonable risk and has been injured as a result of the voluntary acceptance of such risk. The basis for this doctrine is that the fire service organization should not be liable for injuries which the fire fighter knowingly and voluntarily accepted as part of the job. Generally, risks that the employer has no greater knowledge of and which are not unusual in nature should not serve as the basis for the fire service organizations liability. For example, when an individual joins a fire service organization, the individual assumes the risk of working in burning buildings as an inherent part of the job responsibilities. If the individual is injured while fighting a fire inside a building, the individual will be able to acquire coverage under most workers compensation laws but could be barred under the Doctrine of Assumption of the Risk for other actions sounding in tort.

FELLOW SERVANT RULE

Fire fighters may also be restricted in their ability to recover for injuries caused by fellow fire fighters. Under the "Fellow Servant Rule," when an individual is injured by the acts of a co-worker in the course of their employment, the co-worker and fire service organization cannot be held liable. It should be noted that this Rule is effective in only a few states and, in general, has received a substantial amount of criticism.

DEFECTIVE PRODUCTS

With the expanding technology being used by fire service organizations today, the potential of encountering a defective product which could injure a fire fighter has increased significantly. Today, virtually all states have adopted the rule of *MacPherson v. Buick Motor Company*,[2] which held the manufacturer of a defective product liable in negligence to a person injured by the product, despite the absence of privity between the parties. Additionally, most jurisdictions have adopted the warranty theory that manufacturers of defective products can be held strictly liable in tort.[3] Many states have developed their own laws with regard to defective products[4] usually based upon the Restatement (Second) of Torts §402A.

Manufacturers, suppliers, assemblers, and component parts manufacturers can all be liable for a defective product. Individuals protected by most product liability laws include the actual user of the product, consumers, bystanders, and even rescuers if the defective product threatened or caused injury during the rescue of the imperilled individual.

Under the strict liability laws, foreseeability of harm by the manufacturer is not required. Strict liability can apply even if the defec-

---

[2]Supra.

[3]See, *Greenman v. Yuba Power Products, Inc.* 59 Cal.2d 67, 27 Cal. Rptr. 697, 377 P.2d 897 (1963).

[4]See, as example, Kentucky Product Liability Act, KRS 411.300 et. seq.

tive product was being misused at the time of the injury, provided the misuse was foreseeable.[5] Additionally, where the manufacturer improperly designed the product, there is no defense that the design defect was obvious to the consumer.[6] However, where a product has unavoidably unsafe characteristics, such as certain drugs, a warning may be required. For example, the Pasteur treatment for rabies can lead to serious injury and damaging consequences when injected. Since the disease itself has the potential of a painful death, both the marketing and use of the vaccine was justified notwithstanding the high degree of risk.[7] Under this principal, most courts hold that manufacturers can rely on warnings, such as labels for drugs, identifying the risks of the product. However, failure to communicate the risks by third parties, such as physicians with prescription drugs, may result in strict liability.[8]

Strict liability may be applied if the product is unreasonably dangerous and the manufacturer fails to provide proper warnings of the dangers or proper directions as to the use of the product.[9] The duty to warn may be present even where the use by the consumer of the product is abnormal if the abnormal use is a foreseeable use.[10]

Circumstantial evidence may be used to prove strict liability just as it can be used to prove negligence. The circumstantial evidence must be sufficient to warrant the inference that the defective condition existed and was linked to the individual's injury.[11] For example, if a fire fighter was injured when his turnout gear melted and

---

[5] *Moran v. Fabrege, Inc.* 273 Md. 538, 332 A.2d 11 (1975).

[6]*Palmer v. Massey-Ferguson, Inc.*, 3 Wash. App. 508 (1970)("The law, we think, ought to discourage misdesign rather than encourage it in its obvious form.")

[7]See also, *Hines v. St. Joseph's Hospital*, 86 N.M. 763, 527 P.2d 1075 (1974) (No process to determine hepatitis virus in blood supply at that time).

[8] See, *Davis v. Wyeth Labs., Inc.*, 399 F.2d 121 (9th Cir. 1968) (live polio vaccine – strict liability found).

[9] Id.

[10]See, *Spruill v. Boyle – Midway, Inc.*, 308 F.2d 79 (4th Cir. 1962)(14th month old infant died from ingesting furniture polish – Failure to warn of toxic nature).

[11]See, *Elmore v. American Motors Corp*, supra.; Briner v. General Motors Corp. 461 S.W.2d 99 (Ky. 1970); *Perkins v. Trailco Mfg. & Sales Co.*, 613 S.W.2d 855 (Ky. 1981).

caught fire, the proof that the manufacturer produced the turnout gear and the turnout gear melted at a temperature in which the gear was supposed to provide protection should be sufficient to achieve this *causal nexus*.

Two of the often used defenses to a strict liability claim by fire fighters are assumption of risk and contributory negligence.

Assumption of the risk generally bars recovery when an individual who knew the danger in the situation but nevertheless voluntarily accepted the risk. An "Assumption of the Risk" defense is basically an argument that the individual unreasonably accepted the risk and this acceptance of the risk should protect the defendant. The general test for this defense is that the injured individual must have been aware of the defect in the product and not merely that a reasonable person should have been aware of the defect.[12] This is a subjective test and the jury is not required to accept the injured individual's testimony. (Note: Other factors such as age, experience, knowledge, understanding, obviousness of defect, and danger the defect poses can also be taken into consideration.)

Assumption of the Risk clauses are often used in contracts for employment. In most circumstances, the fire service organization or employer cannot be used where the fire fighter was placed in a position of danger due to the fire service organization's negligence. The fire service organization or employer must be free of negligence in order for the Assumption of Risk defense normally to be applicable.[13]

Contributory negligence is usually not a defense to strict liability with the exception of when the injured individual voluntarily and unreasonably proceeds to encounter a known danger.[14] In comparative negligence jurisdictions, an injured individual found guilty of contributory negligence can have the amount of the damages reduced by the respective percentage of guilt.[15]

---

[12]See, *Williams v. Brown Mfg. Co.*, 45 Ill.2d 418, 261 N.E.2d 305 (1970).

[13]See, *Jackson v. City of Kansas City*, 235 Kan. 278, 680 P.2d 877 (Kan. 1984).

[14]Restatement (Second) of Torts § 402(A), Comment n.

Fire service organizations and fire fighters should be aware of the potential dangers of the products that they use on a daily basis. Close scrutiny of warranties, product guarantees, and product usage should be part of any/all purchasing analysis. (Congress is currently debating changes to the product liability laws, which may limit the amount of recovery and provide other modifications to this area of the law.)

OSHA AND OTHER GOVERNMENTAL AGENCIES

In 1970, Congress legislated the Occupational Safety and Health Act (known as the "OSH Act") into law. This Act covers employment in every state, the District of Columbia, Puerto Rico, and all other American territories. The OSH Act encompasses an estimated 5 million work places and over 75 million employees. The OSH Act does not apply to state and federal agencies exercising authority to prescribe or enforce regulations or standards regarding the safety and health of these employees. Additionally, individuals states, such as Kentucky, Iowa, and California, may elect to establish their own agency to enforce safety and health within the boundaries of their state and petition the Occupational Safety and Health Administration to develop a separate and independent regulatory scheme for their individual state. All state plans must have a regulatory scheme which is at least as stringent as the federal regulatory scheme under the OSH Act.

The OSH Act provided for the establishment of three independent branches under which the OSH Act would be managed and regulated. The Occupational Safety and Health Administration (hereinafter "OSHA"), located in the Department of Labor, receives the most notoriety and is the enforcement branch under the OSH Act. The National Institute of Occupation Health and

---

[15]See, *Daly v. General Motors Corp.*, 144 Cal. Rptr. 380, 575 P.2d 1162 (1978).

Safety (hereinafter "NIOSH"), provides the research and testing on matters of safety and health and is located in the Department of Health and Human Services. The Occupational Safety and Health Review Commission (hereinafter "OSHRC") is the independent judicial branch and provides one of the administrative appeal processes under the OSH Act.

Among the many requirements provided under the OSH Act and the standards established since 1970, two baseline standards are of utmost importance. Under Section 5(a)(1), also known as the "general duty clause," every employer is required to maintain its place of employment free from recognized hazards that are causing or are likely to cause death or serious physical harm to employees. The second baseline standard is Section 5(a)(2) which requires the employer to comply with all promulgated OSHA standards.

The OSH Act established three basic ways in which an OSHA standard could be promulgated. Under Section 6(a), the Secretary of Labor was authorized to adopt national consensus standards and establish federal standards without the rule-making procedures normally required under the OSH Act. This authority granted to OSHA ended on April 27, 1973. The second method by which OSHA may promulgate a standard is under Section 6(b) which establishes the procedures to be followed in modifying, revoking, or issuing new standards. This is the normal method by which OSHA establishes new standards today. The third, though seldom used, method was provided under the OSH Act and is the emergency temporary standard under Section 6(c). Emergency Temporary Standards may be issued by the Secretary of Labor if employees are subject to grave danger from exposure to substances or agents known to be toxic or physically harmful and a standard is necessary to protect employees from harm. These standards are effective immediately upon publication in the Federal Register and are in effect for a period not to exceed six months.

The enforcement function under the OSH Act is provided to OSHA. OSHA compliance officers are empowered by Section 8(a) to inspect any work placed covered by the OSH Act, subject to the limitations set forth under *Marshall v. Barlow*.[16] An OSHA compliance officer is required to present his/her credentials to the owner or manager prior to proceeding on an inspection tour and the owner or manager has the right to accompany the compliance officer during the inspection tour. Employees or union representatives also have the right to accompany the compliance officer on his/her inspection tour. After the inspection, the compliance officer will hold a "closing conference" with the employer and employee representatives at which time the safety and health conditions and potential citations or violations will be discussed. Most compliance officers do not have the authority to issue "on the spot" citations but must confer with the regional or area director before issuing citations. Compliance officers do have the authority to shut down an operation which is life threatening or which places employees in a position of imminent danger. As will be discussed, employers should be prepared for an OSHA inspection and develop a program to ensure all rights and responsibilities provided under the OSH Act are being complied with by all affected parties.

Following the closing conference, the compliance officer is required to issue a report to the area or regional director. The area or regional director usually decides whether to issue a citation and assesses any penalty for the alleged violation. Additionally, the area or regional director sets the date for compliance or abatement for each of the alleged violations. If a citation is issued by the area or regional director, this notice is mailed to the employer as soon as possible after the inspection, but in no event can this notification be more than six months after the alleged violation occurred. Citations must be in writing and must describe with particularity the alleged violation, the relevant standard and regulation, and the date of the alleged violation.

The OSH Act provides for a wide range of penalties from a simple notice with no fine through criminal prosecution. Violations are

---

[16]Supra.

categorized and penalties may be assessed in the manner shown in Table 6.1.

Each alleged violation will be categorized and the appropriate fine issued by the area or regional director. Please note that each citation is separate and may carry with it a monetary fine. OSHA has defined a serious violations as "an infraction in which there is a substantial probability that death or serious harm could result... unless the employer did not or could not with the exercise of reasonable diligence, know of the presence of the violation."[17]

The greatest monetary liability which currently exists is for the repeat violations, willful violations, and failure to abate cited violations. A repeat violation is simply being cited a second time for a violation which had previously been cited by a compliance officer. OSHA maintains records of all violations and is required to check for repeat violations after each inspection. A willful violation is the employer's purposeful or negligent failure to correct a known deficiency. This type of violation, in addition to carrying a large monetary fine, opens the employer to the potential of criminal sanctions under the OSH Act or through state criminal statutes if an employee is injured or killed as a direct result of the willful violation. Failure to abate a cited violation results in the greatest cumulative mone-

---

[17]See, 29 U.S.C. § 666.

TABLE 6.1   Violations and Penalties

	Old Penalty Schedule	New Penalty Schedule (1990)
De Minimis Notice	$0	$0
Non-serious	$0 to $1,000	$0 to $7,000
Serious	$1 to $1,000	$0 to $7,000
Repeat	$0 to $10,000	$0 to $70,000
Willful	$0 to $10,000	new minimum, $25,000 maximum, $70,000
Failure to Abate Notice	$0 to $1,000 per day	$0 to $7,000 per day
New Posting Penalty		$0 to $7,000

tary liability of all violations. OSHA may assess a penalty up to $1,000.00 per day, per violation, for each day in which a cited violation is not brought into compliance. In assessing monetary penalties, the area or regional director should take into account the good faith of the employer, the gravity of the violation, the employer's past history of compliance, and the size of the employer.

The OSH Act also carries the potential of criminal prosecution of management for work place safety and health violations. Since the establishment of OSHA in 1970 through 1989, there have only been 42 cases referred for criminal prosecution of which approximately one-third were ultimately prosecuted. This seldom-used penalty under the OSH Act is becoming more frequently used because of the scrutiny of the press and public and the use of state prosecutions of employers for work place deaths. In Illinois, Texas, and several other states, state prosecutors have successfully brought criminal charges against members of management for work place deaths.[18] At this time, the current status of the law is that the OSH Act does not preempt state and local prosecution in the areas which are covered by the OSH Act.

Under the OSH Act, the employer is required to post the alleged violation within its facility. The employer, union, or individual employee has 15 working days in which to file a notice of contest with the OSHA office. The Secretary of Labor is required to forward any/all notice of contest to the Occupational Safety and Health Review Commission (OSHRC). The employer has the right to appeal the final order of the OSHRC. The employer may file a petition for review in the United States Court of Appeals for the circuit in which the alleged violation occurred or in the United States Court of Appeals for the District of Columbia Circuit. The employer must file this petition within 60 days of the final order.

Senator Metzenbaum of Ohio and Congressman Lantos of California first introduced a bill called the *OSHA Criminal Penalty*

---

[18]See, *People v. O'Neil* 135 Ill. App. 1091, 90 Ill. Dec. 689, 482 N.E. 2nd 688 (1985); *People v. Chicago Magnet Wire Corp.* 126 Ill. 2nd 356, 129 Ill. Dec. 517, 57 U.S.L.W. 2460, 534 N.E. 2nd 962 (1989).

Reform Act into Congress in 1990. This proposed bill would substantially modify the OSH Act and include fire service organizations within the jurisdiction of OSHA. This bill has been re-introduced in the Senate (S 445) by Senator Metzenbaum (D., Oh.) and Senator James Jefford (R., Vt.) and in the House (HR 192) by Representative Lantos (D., Cal.) with substantial support from OSHA. Additionally, Representative Charles Schumer (D., NY) has introduced the *Workplace Protection Act of 1991* (HR 549).

Under the provisions of Senate Bill 575 and House Bill 1280 currently before Congress, [known as the "Comprehensive Occupational Safety and Health Reform Act (COSHRA)"], the jurisdiction of OSHA may be broadened to include all public sector employers and thus most fire service organizations. Additionally, this significant legislation proposes to increase the potential fines for willful violations of the OSHA standards which result in a fatality of up to $250,000.00 for an individual (i.e., the fire chief or incident commander) and up to $500,000.00 for a corporation or organization.[19] If a willful violation of an OSHA standard results in the death of an employee, criminal penalties would be increased from the currently level of 6 month to 10 years for the first offense and 20 years for the second offense. If a willful violation occurs and serious bodily injury is incurred by an employee, a fine of up to $250,000.00 and 5 years imprisonment can be imposed on the individual and the corporation or organization will be prohibited from paying the monetary penalties on behalf of the officer.

Under Title V of the COSHRA, supervisors, line officers, and other management officials who violate the Occupational Safety and Health Act or who direct someone else under their control to violate the OSH Act or standards, if the violation causes the death of or serious bodily injury to an employee, the supervisor or line officer can be held criminally liable for these actions.

---

[19]Under the Federal Sentencing Reform Act of 1978, any misdemeanor involving violation of federal law which results in a fatality is punishable by a minimum fine of $250,000.00 for an individual and $500,000.00 for a corporation. The U.S. Justice Department has taken the position that these fines apply to willful OSHA violations resulting in an employee death.

Under Title I of the COSHRA, all employers will be required to establish and carry out a written safety and health program. At a minimum, the written program may be required to contain:

- methods and procedures for identifying, evaluating, and documenting safety and health hazards
- methods and procedures for investigating work-related illnesses, injuries, and deaths
- methods and procedures for correcting safety and health hazards identified through implementation of the program
- methods and procedures for providing emergency response first-aid and other occupational safety and health services
- methods and procedures for employee participation in the implementation of the program, including mandatory joint safety and health committees at most worksites with eleven or more employees
- designation of one or more employer representatives who have qualifications and responsibility to identify and initiate corrective action with regard to safety and health hazards

The above, as well as additional requirements, is just a small portion of the many significant changes which could directly affect a fire service organization if and when COSHRA becomes law. The rules will be changing, the risks will increase, and the penalties will be much higher.[20]

All indicators point to a definite increase in criminal sanctions under the OSH Act. At a recent Senate subcommittee hearing, Mr. Gerald F. Scannell, head of OSHA, stated the Bush Administration supports the idea of upgrading willful work place deaths from misdemeanors to felonies, however, the Administration opposes the expansion of criminal sanctions beyond fatality cases.

---

[20]At the writing of this text, the COSHRA bill is still pending in committee. The author does not anticipate this proposed legislation to become law prior to 1996, if at all.

Since the enactment of the OSH Act, four (4) major areas of litigation have arisen. The first area of potential conflict is in the area of promulgation of standards. The OSH Act does not describe the specific hazards to be regulated or the methods to reduce or eliminate the specific hazards to be regulated or the methods to reduce or eliminate the hazards. The responsibility for all rule-making activities was delegated to the Secretary of Labor. As discussed above, there are three methods by which OSHA standards may be promulgated. This methodology has created numerous technical, policy, and legal issues arising out of each of these three rule-making methods.

The second major area of conflict and litigation is the inspection procedures by OSHA. The threat of an OSHA inspection and the possibility of a citation carrying with it the potential of fines or criminal sanctions is intended to prod employers to meet their pre-inspection duty to comply with OSHA standards. For employers who disregard the OSH Act and the applicable standards, the enforcement process involves the government's effort to penalize prior dereliction and to require the prompt correction of identified and existing hazards in the work place.

OSHA inspections can be basically divided into four categories and have been assigned the following priority: (1) imminent dangers, (2) fatality and catastrophe investigations, (3) inspections of employee complaints, (4) regional program or "target" inspections which concentrates on high hazard industries with a large number of employees.

Most OSHA inspections begin with a knock on the door and an OSHA inspector showing his/her credentials and stating the purpose of their inspection. Prior to 1978, the OSH Act permitted OSHA inspectors to gain immediate and unrestrained access to any work place covered under the OSH Act. In a case called *Marshall v. Barlow*,[21] the Supreme Court held an employer may require that an OSHA inspector to show probable cause and acquire a valid search

---

[21]436 U.S. 307 (1978)

warrant before gaining access to an employer's property (See Appendix C for the full text of this case). There are three exceptions to an employer's ability to require an OSHA inspector to acquire a warrant before gaining access to the work place. In *Camara v. Municipal Court*,[22] and *See v. Seattle*,[23] a warrant is not required if the employer consents to the inspection, if there is an open view of the hazard by the OSHA inspector, or if there is an emergency. Since these cases, the instances of litigation regarding OSHA access to an employer's work place and the scope of the inspection, especially complaint inspections, has increased substantially.

To avoid litigation in this area, the Fire service organizations must decide beforehand the company's position and policy as far as requiring an OSHA inspector to acquire a warrant and/or the scope of permissible inspection. In no way will an employer be able to completely bar OSHA or the applicable state plan inspectors from acquiring access to the work place, but the employer should plan in advance and discuss the potential approaches which are appropriate for the individual company prior to the inspection process to ensure all rights granted under the OSH Act are protected and all responsibilities are fulfilled.

A recent decision in *Electromation, Inc. v. NLRB*,[24] has also caused some concern to fire service organizations in the area of safety and other committees. This decision by the National Labor Relations Board and supported by the Seventh Circuit Court of Appeals found that a committee which was improperly structures and dominated by a company or organization could constitute an unfair labor practice under the National Labor Relations Act. This decision, although narrow in scope, may affect the structure of some committees within the fire service however most fire service

---

[22] 387 U.S. 523 (1967)

[23] 387 U.S. 541 (1967).

[24] 309 NLRB No. 163 (1992)

organizations utilizing safety and other committees will not be affected by this decision.

First and foremost, it must be noted that there is no substitute for a properly designed, effectively managed, and an "in compliance" safety program. An effective safety program will not only achieve the goals set forth by Congress under the OSH Act, but also return dividends in the areas of reduced workers compensation and other insurance costs, improved employee morale, and numerous other benefits for the employers and the employees. The OSHA standards and guidelines should be the "bare bones" minimum under which an employer should design his/her safety and health programs. Employers should take the initiative to go far beyond the basic compliance areas to develop new and innovative ideas and programs to promote the safety and health of their employees while insuring complete and total compliance with the basic safety standards required in the OSHA standards and guidelines.

With any OSHA or state plan compliance inspection, the employer must be prepared before the compliance officer is standing on the door step! After the employer decides which tact the company will adopt and follow, specific guidelines should be drafted and the appropriate managerial personnel should be properly trained to manage each stage of the inspection process. Specific forms should be developed for the management personnel to use in gathering and documenting all information before, during, and after the inspection. A notification procedure and decision making hierarchy for the appropriate members of management should be established. The appropriate management personnel should be equipped with cameras, tape recording devices, video-tape recorders, and all other required testing equipment in order to be able to acquire identical information which the compliance officer is acquiring during the inspection. It is vitally important to document everything before, during, and after the inspection!

The third major area of litigation is the contest of citations. Once the area or regional director issues a citation, the employer has 15 days to contest the citation in writing to the regional or area director. The employer should review and evaluate the citation closely and ensure all information is accurate and the appropriate standard is cited for the alleged violation. The first stage of appeal is normally to the area or regional director. An informal hearing may be requested to discuss the alleged violation, the standard section cited, and/or the fines. The regional or area director usually has the authority to vacate any citation and usually can reduce the fine by as much as 50%. The good faith of the employer, past history, effectiveness of the employer's overall safety and health program, and other related areas are usually considered by the area or regional director. Good documentation and valid, effective arguments can assist the area or regional director in settling the citation at this stage.

The next step is a hearing with the administrative law judge (ALJ) for the OSHRC. The ALJ has the authority to vacate any citation or penalty. All ALJ decisions are reviewed by the OSHRC. As stated above, proper documentation acquired during the inspection itself is a must at this stage of the appeal process.

The employer may request appeal to the OSHRC, but if the OSHRC does not render a decision within 30 days, the ALJ's decision stands. The other route of appeal is through the U.S. Court of Appeals for the circuit in which the alleged violation occurred or the D.C. Circuit Court of Appeals in Washington, D.C. Most employers are more comfortable in the federal court system near the employer's place of business which provides easy access to witnesses, less time away from the work place for management, and other peripheral benefits. Remember, this is a *trial de novo*. The employer must gather evidence to support his/her case in the same manner as any other trial thus the documentation assembled during the inspection now becomes the keystone of the employer's appeal.

The fourth area of potential litigation is the blossoming arena of individual and corporate criminal liability for members of management for non-compliance with work place safety and health standards and work place injuries and deaths. With the increased use of criminal sanctions by OSHA, the advent of individual states taking the initiative to prosecute employers for work place deaths as in the *Film Recovery* and *Chicago Magnet Wire* cases, and the possibility of reformation under the *OSHA Criminal Penalty Reform Act* and/ or *Workplace Protection Act of 1991*, the safety and health area will again become an area of controversy.

In conclusion, litigation under the Occupational Safety and Health Act seldom makes the front page of your local newspaper but occurs regularly throughout the administrative process and in the federal courts. With the new penalty increase and the increased use of the criminal sanctions, the Occupational Safety and Health Act will become a "hotbed" of litigation in the 1990s.

SELECTED CASES

(This case has been edited for the purposes of this text.)

Barger v. Mayor & City Council of Baltimore
616 F.2d 730

Warren Barger and Marion Iwancio are former members of the fire department of the City of Baltimore. Barger and Iwancio worked in the Marine Division of the department as crew members of the city's fireboats. They began to suffer from hearing loss which had stemmed from their prolonged exposure to the loud noise emitted by the engines of the diesel fireboats. Barger and Iwancio were forced to retire because of their hearing impairments and received special disability pensions from the city.

Subsequently, Barger and Iwancio brought this suit under the Jones Act and general maritime law to recover damages for their loss of hearing, based upon the city's alleged negligence in the operation of the fireboats with the noisy engines. The jury found that the city was negligent and ruled in favor of Barger and Iwancio under both the Jones Act and general maritime law. Barger was awarded $153,000 and Iwancio received $112,500.

The city appealed principally on the basis of the district judge's instructions to the jury concerning negligence *per se*. The district judge instructed the jury that the city had been subject to the regulations of the Occupational Safety and Health Administration (OSHA) since 1971 and that if the city failed to comply with the OSHA noise regulation in the operation of its fireboats, it would be guilty of negligence *per se*.

The city objected to the instructions on the basis that the city of Baltimore became subject to OSHA regulations only in 1973, not as early as 1971 as the trial judge instructed. The city also objected that the OSHA noise regulation does not apply to working conditions of Jones Act seamen and that therefore, the trial judge committed error by including any negligence per se instruction based on the OSHA noise standards.

The Court of Appeals concluded that while the judge did misstate the date upon which OSHA regulations began to apply to the city, this two year discrepancy was of minor significance and that the jury only applied a small part of the period during which the firemen were exposed to the loud noises. The Court also ruled that the district judge did err when he included a negligence per se instruction based upon the OSHA noise regulation but that the city could not obtain a reversal of the jury's verdict based upon this error. When the district judge presented the city with his proposed instructions, the city made no objection based upon the inapplicability of the OSHA noise regulation to Jones Act seamen. In fact, the city expressly agreed to the district judge's proposed instructions before he presented them to the jury.

Affirmed.

U.S. DEPARTMENT OF LABOR

OCCUPATIONAL SAFETY AND HEALTH ADMINISTRATION

OMB No. 044RJ449

	For Official Use Only		
Area	Date Received		Time
Region	Received By		

COMPLAINT

This form is provided for the assistance of any complainant and is not intended to constitute the exclusive means by which a complaint may be registered with the U.S. Department of Labor.

The undersigned (check one)

☐ Employee   ☐ Representative of employees   ☐ Other (specify) _____

believes that a violation at the following place of employment of an occupational safety or health standard exists which is a job safety or health hazard.

Does this hazard(s) immediately threaten death or serious physical harm?   ☐ Yes   ☐ No

Employer's Name _____

Address
(Street _____   Telephone _____
(
(City _____   State _____   Zip Code _____

1. Kind of business _____

2. Specify the particular building or worksite where the alleged violation is located, including address. _____

3. Specify the name and phone number of employer's agent(s) in charge. _____

4. Describe briefly the hazard which exists there including the approximate number of employees exposed to or threatened by such hazard.

_____
_____
_____
_____
_____
_____
_____

(Continue on reverse side if necessary)

Sec. 8(f)(1) of the Williams-Steiger Occupational Safety and Health Act, 29 U.S.C. 651, provides as follows: Any employees or representative of employees who believe that a violation of a safety or health standard exists that threatens physical harm, or that an imminent danger exists, may request an inspection by giving notice to the Secretary or his authorized representative of such violation or danger. Any such notice shall be reduced to writing, shall set forth with reasonable particularity the grounds for the notice, and shall be signed by the employees or representative of employees, and a copy shall be provided the employer or his agent no later than at the time of inspection, except that, upon request of the person giving such notice, his name and the names of individual employees referred to therein shall not appear in such copy or on any record published, released, or made available pursuant to subsection (g) of this section. If upon receipt of such notification the Secretary determines there are reasonable grounds to believe that such violation or danger exists, he shall make a special inspection in accordance with the provisions of this section as soon as practicable, to determine if such violation or danger exists. If the Secretary determines there are no reasonable grounds to believe that a violation or danger exists he shall notify the employees or representative of the employees in writing of such determination.

[E5938]

Form OSHA-7
Rev. Jan. 1972

5. List by number and/or name the particular standard *(or standards)* issued by the Department of Labor which you claim has been violated, if known.

_____

_____

_____

6. (a) To your knowledge has this violation been considered previously by any Government agency? _____

(b) If so, please state the name of the agency _____

(c) and, the approximate date it was so considered. _____

7. (a) Is this complaint, or a complaint alleging a similar violation, being filed with any other Government agency?

(b) If so, give the name and address of each. _____

8. (a) To your knowledge, has this violation been the subject of any union/management grievance or have you *(or anyone you know)* otherwise called it to the attention of, or discussed it with, the employer or any representative thereof? _____

(b) If so, please give the results thereof, including any efforts by management to correct the violation. ___

9. Please indicate your desire:

☐ I do not want my name revealed to the employer.
☐ My name may be revealed to the employer.

Continue Item 4 here, if additional space is needed.

_____

_____

_____

_____

_____

_____

COMPLAINANT'S NAME

Signature _____   *Date* _____

Typed or Printed Name _____

Address (*(Street* _____   *Telephone* _____

  (City _____ *State* _____ *Zip Code* _____

If you are a representative of employees, state the name of your organization _____

U.S. DEPARTMENT OF LABOR

Occupational Safety and Health Administration

MATERIAL SAFETY DATA SHEET

Form Approved
OMB No. 44–R1987

Required under USDL Safety and Health Regulations for Ship Repairing,
Shipbuilding, and Shipbreaking (29 CFR 1915, 1916, 1917)

SECTION I

MANUFACTURER'S NAME	EMERGENCY TELEPHONE NO.
ADDRESS (Number, Street, City, State and ZIP Code)	

CHEMICAL NAME AND SYNONYMS Benzene *	TRADE NAME AND SYNONYMS Benzol
CHEMICAL FAMILY Organic	FORMULA C_6H_6

SECTION II—HAZARDOUS INGREDIENTS

PAINTS, PRESERVATIVES, & SOLVENTS	%	TLV (Units)	ALLOYS AND METALLIC COATINGS	%	TLV (Units)
PIGMENTS			BASE METAL		
CATALYST			ALLOYS		
VEHICLE			METALLIC COATINGS		
SOLVENTS			FILLER METAL PLUS COATING OR CORE FLUX		
ADDITIVES			OTHERS		
OTHERS					

HAZARD MIXTURES OF OTHER LIQUIDS, SOLIDS, OR GASES	%	TLV (Units)

SECTION III—PHYSICAL DATA

BOILING POINT (° F.)	176	SPECIFIC GRAVITY (H_2O=1)	0.88
VAPOR PRESSURE (mm Hg.) @ 26.1 ° C.	100mm	PERCENT, VOLATILE BY VOLUME (%)	100
VAPOR DENSITY (AIR=1)	2.77	EVAPORATION RATE (= 1)	
SOLUBILITY IN WATER	sl. Sol		

APPEARANCE AND ODOR Clear colorless liquid, characteristic odor.

SECTION IV—FIRE AND EXPLOSION HAZARD DATA

FLASH POINT (Method used) 12 F (Closed Cup)	FLAMMABLE LIMITS	Lel 1.3%	Uel 7.1%

EXTINGUISHING MEDIA
Foam is best, CO_2 and dry chemical

SPECIAL FIRE FIGHTING PROCEDURES
Wear self-contained breathing apparatus.

UNUSUAL FIRE AND EXPLOSION HAZARDS
When fighting fire use necessary protective
equipment and breathing apparatus.

* See note in Section No. IX—Special Precautions

PAGE (1)   (Continued on reverse side)   [E5940]   Form OSHA–20

SECTION V—HEALTH HAZARD DATA

THRESHOLD LIMIT VALUE
25 mg/M^3 Air. LD_{50} ORAL (RAT) = 3400mg/Kg.

EFFECTS OF OVEREXPOSURE
Dizziness, headache, breathlessness, and excitement followed by mental confusion

and hysterical symptoms.

EMERGENCY AND FIRST AID PROCEDURES
1.) Remove from exposure 2.) Remove contaminated clothing 3.) If in eyes, flush with water for 15

minutes. Call physician. 4.) Internally—induce vomiting, call doctor. Do not give anything by mouth

if unconscious 5.) If inhalation—lie down, keep warm, call physician.

SECTION VI—REACTIVITY DATA

STABILITY	UNSTABLE		CONDITIONS TO AVOID
	STABLE	x	

INCOMPATABILITY (Materials to avoid)
Oxidizing materials

HAZARDOUS DECOMPOSITION PRODUCTS
Yes—All types of products in case of fire.

HAZARDOUS POLYMERIZATION	MAY OCCUR		CONDITIONS TO AVOID
	WILL NOT OCCUR	x	

SECTION VII—SPILL OR LEAK PROCEDURES

STEPS TO BE TAKEN IN CASE MATERIAL IS RELEASED OR SPILLED
Small Quantities—absorb on paper. Evaporate in hood. Wear proper safety equipment.

WASTE DISPOSAL METHOD
Atomize into an incinerator

SECTION VIII—SPECIAL PROTECTION INFORMATION

RESPIRATORY PROTECTION (Specify type)
Handle in hood or use self-contained breathing apparatus.

VENTILATION	LOCAL EXHAUST		SPECIAL
	MECHANICAL (General) X (Sparkproof Fans)		OTHER

PROTECTIVE GLOVES Synthetic rubber	EYE PROTECTION Cup type or rubber framed goggles.

OTHER PROTECTIVE EQUIPMENT

SECTION IX—SPECIAL PRECAUTIONS

PRECAUTIONS TO BE TAKEN IN HANDLING AND STORING
Outside or detached storage is preferable.

OTHER PRECAUTIONS
Inside storage should be in a standard flammable liquids storage room or cabinet.

NOTE: This item is listed as a cancer-suspect agent by OSHA as of 6/2/80.

(R.S. & S.) Labor & Employ. Law ACB—7
     Stat.Supp.

7

N.F.P.A., B.O.C.A., N.E.C., and Others

In the fire service, the acronyms NFPA, BOCA, NEC, and others are commonly used to refer to the recommendations and guidance provided by these various codes and/or organizations. Fire service organizations should be aware that these codes and rules may be advisory in nature or can be compulsory depending on the type of code and whether the particular code has been adopted by another governmental entity or jurisdiction which mandates compliance. For example, the National Electrical Code has been adopted by many jurisdictions as the standard or "law" which electrical wiring, installation, and other electrical practices must comply in order to be approved. Additionally, OSHA and most state-plan states have adopted several National Electrical Code sections and codified these codes within the OSHA standards. Fire service personnel should be aware as to the various codes which directly or indirectly affect their organization and the status as to whether the code is advisory or mandatory within their jurisdiction.

Fire protection and the functions of a fire organizations have been studied for many years by various governmental and private organizations. As a result of these studies, most jurisdictions have developed specific laws to safeguard the public and the fire fighter in the form of statutes, ordinances, and private industry codes such as OSHA. As a general rule, these laws were developed as a reaction by the governmental body after a catastrophic fire had occurred in order to prevent recurrence. Virtually every jurisdiction has these types of statutes or ordinances on the books and these laws are normally focused on the occupancy, structure, and other relevant circumstances of the particular jurisdiction. Fire service personnel should be familiar with the specific codes and ordinances within their jurisdiction. Given the fact that these laws are usually mandatory in nature, all within the particular jurisdiction are required to comply with the law.

Distinctions must be drawn between laws, statutes, codes, or standards which mandate compliance and codes or standards which are advisory in nature. The "lines" between whether a code is mandatory or voluntary can often be blurred and this distinction often can only be made by considering to which jurisdiction the code is being applied. This can often be confusing. However, there are a few keys which can assist you in making this distinction:

1. statutes, ordinances, public laws, or administrative regulations are usually mandatory in nature
2. codes, standards, or regulations developed by associations or private entities are usually advisory in nature
3. codes, standards, or regulations developed by associations or private entities can and often are adopted by governmental entities. Look for the source of the advisory standard or code and see if the code or standard has been adopted or referred to in a mandatory standard

Probably the must common acronym used in the fire service is NFPA. NFPA stands for the National Fire Protection Association. The National Fire Protection Association has published a handbook on various fire protection topics since 1896 and is often referred to as the "Bible" for fire service organizations.[1] Within the NFPA Handbook, the authors cover various topics ranging from the

---

[1]NFPA Handbook (Seventeenth Edition) at p. 1.

Basics of Fire and Fire Service to Fire Suppression. The NFPA handbook is often considered the baseline or level of expected performance or behavior which all fire service organizations should maintain. However, the NFPA handbook and the sections contained therein, such as NFPA 1500, are advisory in nature unless codified or adopted by the specific jurisdiction.

Of particular note to fire fighters should be NFPA 1500, 1582, and 1583 as proposed. NFPA 1500 deals with the physical requirements of a fire fighter and NFPA 1582 is the standard on medical requirements for fire fighters. These particular NFPA standards have been controversial in the past and tend to cause fire service organizations substantial problems in achieving and maintaining compliance. NFPA 1583 as proposed (Recommended Practice for Physical Performance Testing of Fire Fighters) should be reviewed by fire service organizations. These particular NFPA standards have application as related to the Americans with Disabilities Act (see Chapter 8) especially when dealing with fire fighters who are HIV Positive or have other disabilities.

Given the extensive nature of the NFPA Handbook, it is not unusual to find that specific codes or sections have been adopted, in whole or in part, as the official fire or safety code of a municipality, a state, or even federal governmental agencies. The NFPA Codes, as with most standards or codes, can be extremely broad in nature or can be very narrowly drawn especially in highly technical or specialized areas.

Another often used acronym in the fire service is ANSI. ANSI stands for the American National Standards Institute. As with the NFPA standards, ANSI standards are also advisory in nature. However, a substantial number of the ANSI standards have been adopted within the Occupational Safety and Health standards and codified within other specific federal, state, and local codes.[2] ANSI standards tend to be broad in nature encompassing a vast number

---

[2] 29 C.F.R. § 1910.

of topics and tend to be a forerunner in advanced safety and health issues such as ergonomics.

A code which has been adopted in whole by a substantial number of federal, state, and local jurisdictions is the Life Safety Code. The Life Safety Code is often used by fire service organizations when evaluating construction of new or existing buildings, evaluating exits and egress, and similar specific areas. Similar to the Life Safety Code is the BOCA codes. BOCA stands for Building Officials and Code Administrators basic fire prevention code and is primarily focused on building standards related to fire safety. As with the Life Safety Code, the entire BOCA codes have been adopted by a substantial number of federal, state and local jurisdictions.

The above are but a few of the various codes and regulations which have been developed by governmental agencies and private sector associations. In researching the applicable standard or regulation as applied to the circumstances involving your fire service organization, please evaluate as to whether the standard is mandatory or voluntary in nature and whether the applicable code or standard has been adopted in your jurisdiction.

Codes and regulations are often used in civil actions, especially negligence actions. The primary purpose of the use of the code or regulation is to show that either a duty was created for the fire service organization or its agents or to show the standard of care which the fire service organization knew or should have known that they were required to comply with as a matter of law.

To exemplify how a code or standard may be used in a civil action, a negligence action has been selected. Common negligence actions against fire service organizations include such actions as failure to respond to a call, negligent fire fighting strategies or techniques, or injury to a person or property because of the actions or inactions or the fire service organization. In a negligence action, the following elements are usually required to be proven by the plaintiff:

1. Duty—Did the fire service organization owe any duty to the plaintiff? If so, what standard of care did the fire service organization owe to the plaintiff under the particular circumstance?
2. Breach of Duty—Did the fire service organization, by their conduct, violate a prescribed duty of care?
3. Causation—Did the fire service organization's conduct factually bring about actual harm to the plaintiff?
4. Proximate Cause—With the assumption that the factual causal connection between the fire service organization's conduct and the actual harm to the plaintiff, was the fire service organization's conduct the proximate cause of the harm to the plaintiff?
5. Damages—Was the negligence of the fire service organization the cause of the particular damages to the plaintiff?

Standards and regulations are normally admissible into court to show that a duty was created or to show the standard of care. Usually the standard or regulation is submitted into evidence through an expert witness or through agreement (stipulation) between the parties. In many circumstances, the issue as to whether the fire service organization had a particular duty to the injured individual(s) is the primary issue in the case. Duty can be an affirmative duty to act or respond or can also be a failure to act or respond when required. Duty can also mean a duty not to act in a negligent or reckless manner. As a general rule, the law does not impose upon individuals the duty to render aid to another person. Moral obligations with respect to responding to particular incidents or accidents does not normally transfer into a legal obligation.[3] However, for fire service organizations, codes and regulations often create an affirmative duty for the fire service organization to respond in an appropriate manner. These codes and regulations often create a special relationship between the fire service organization and individuals residing in the area which they serve. In the event that a

---

[3]*Yania v. Vigan,* 397 Pa. 316, 155 A. 2d 343 (1959) (Defendant not liable for ignoring the pleas of a drowning man.)

duty is created by code, regulation or law, and the fire service organization fails to comply with that duty, the second element of a negligence action is normally achieved, (i.e., breach of duty). Breach of the duty is normally proven by the showing of an affirmative duty to act and the failure of the fire service organization to fulfill that duty.

Upon achievement of the first two hurdles, i.e., duty and breach of duty, the third hurdle in a negligence action is usually proving that the failure to respond to an affirmative duty was the proximate cause of the damages incurred by the plaintiff. Normally, the court uses a variety of tests to see if this causal linkage has been provided. In the event that actual causation exists between the fire service organization's conduct and the plaintiff's harm, no recovery is normally allowed unless the fire service organization's act was the proximate cause of the harm to the plaintiff. The area of proximate cause is normally divided into two categories: (1) harm within the risk; and (2) person within the risk. To ascertain whether there is a harm within the risk, courts use a variety of tests including the "foreseeable consequence" test[4] or the unbroken sequence (in hindsight) test as set forth in *Dellow v. Pearson,* 259 Minn. 452, 107 N.W. 2d 859 (1951).[5] A common problem in negligence actions involving fire service organizations is the area of proximate cause when there is an intervening negligence of a third party. In most jurisdiction, the fire service organization or other defendant will be held liable for all of the consequences brought about by the combination of their negligence and the third party's negligence, if the fire service organization should have foreseen that their negligence would be followed by the negligence of a third party, therefore causing additional injury to the plaintiff.[6] Restatement of

---

[4]*Overseas Tank Ship (U.K.) Limited v. Morts Dock and Engineering Company,* (The Wagon Case), 1 All-Er 404 (1961), overruling *Ree Polemis v. Furness,* Withy and Company, Limited, 3 KB 560 (1921).

[5]This test has been adopted in a minority of jurisdictions.

[6]*Dataloro v. Thomas,* 262 Mass. 383, 160 N.E. 269 (1928) and *Rappaporte v. Nichols,* 31 N.J. 188, 156 A. 2d 1 (1959).

Torts, Second, § 435 sets forth that the courts should use hindsight rather than the foreseeability test in addressing the situation and states as follows:

- If the actor's conduct is a substantial factor in bringing about the harm to another, the fact that the actor neither foresaw nor should have foreseen the extent of the harm or the manner in which it occurs does not prevent him from being liable.
- If the actor's conduct may be held not to be a legal cause of harm to another where, after the event and looking back from the harm of the actor's negligent conduct, it appears to the court highly extraordinary that it should have brought about this harm.

In the event that a fire service organization had a duty to act, and that duty was breached, and the proximate harm to the plaintiff was a result of this breach, the final element in a negligence action is normally damages. Damages can include such things as property damage, punitive damages, attorney's fees, lost wages, pain and suffering, medical expenses, loss of earning capacity, loss of consortium and other damages.

It is important for fire service personnel to maintain competency with regards to the variety of codes and standards and the modifications which are made on a periodic basis. Achieving compliance and maintaining compliance with the codes and standards which may be applicable to your fire service organization in your jurisdiction is imperative in order to avoid potential legal actions.

8

Other Potential Liabilities of the Fire Fighter

A successful lawsuit is the one worn by a policeman.

Robert Frost

PUBLIC OFFICER

Are fire fighters "public officers" or "public employees"? This is the question that has faced many courts and the decisions appear to be split on this issue depending on the jurisdiction. If a fire fighter is determined to be a "public officer," then they are subject to the duties of care imposed on them by law (usually by statute or ordinance) as the holder of this position. In jurisdictions where fire fighters have been found only to be mere "public employees," the fire fighters are only liable for failure to perform their specific duties. This distinction is important in determining the scope of the duty of the fire fighter and the individual potential liability in any given situation.

The scope of authority granted to a fire service organization and to fire fighters is usually determined by statute or ordinance. Most courts have determined that the operations of the fire service would be severely hampered if fire service officials were held accountable by the courts for mistakes or poor judgments in the honest performance of their duties in an emergency situation. Given the fact that often decisions must be made based upon subjective judgment and discretion, most courts tend not to second guess the decision making process so long as the decision was within the scope of authority of the fire service organization and was reasonable under the circumstances.

HAZARDOUS MATERIALS

The activities involved in hazardous materials are fast becoming a major area of concern for fire service organizations. Potential liability in this areas has increased significantly with the passage of numerous environmentally related laws by various governmental agencies such as SARA (EPA) and the Hazard Communication Standard (OSHA). Hazardous material emergencies have also been the source of substantial litigation by individuals (including fire fighters) exposed to chemicals following the release.

In general, fire service organizations must have the authority to operate hazardous materials teams and these teams must be properly trained and prepared prior to responding to a "hazmat" incident. Response absent authority is usually considered *ultra vires* and thus carries the potential for liability for negligence from various sources.

One area often overlooked by fire service organizations is the use of Material Safety Data Sheets (known as MSDS reports) in preparation for response to "hazmat" response activities. MSDS reports are required for all chemicals under the Hazard Communication standard by OSHA.[1] The Material Safety Data Sheet is a document that describes the physical and chemical properties of products, their physical and health hazards, and precautions for safe handling and use. Its function is to provide detailed information on each hazardous chemical, including its potential health effects, its physical and chemical characteristics, and recommendations for appropriate protective measures. Distributors of regulated chemicals are required to furnish their customers with a completed MSDS sheet

---

[1]29 C.F.R. § 1200.

for *each* regulated chemical they ship to you. Customers receiving MSDSs should review them for accuracy and completeness and to ensure that the latest MSDS is on file. A comparison of new and old MSDSs is useful because it may identify those situations where a "new" hazard associated with an existing chemical has been identified, or a new ingredient is included in a currently used product. The following data is required to be included on all MSDSs:

1. chemical identity
2. hazardous ingredients
3. physical and chemical characteristics
4. physical hazards
5. health hazards
6. special precautions, spill, leak, and cleanup procedures
7. control Measures
8. emergency and first-aid procedures
9. responsible party

In addition to the OSHA Hazard Communication Standard,[2] many states have enacted "Right to Know" laws that provide additional access to information. In most jurisdictions, companies having chemicals are required to provide to the fire service organization (or other designated agency) a copy of the facilities Hazard Communication program and all applicable MSDS reports. Fire service organizations should acquire this information and use this important information in hazmat situations.

The proper disposal of hazardous materials following clean-up is another area of potential liability for the fire service organization. Under various environmental laws, specific hazardous materials must be disposed of in specific manners. Fire service organizations should take particular caution in the disposal of chemicals following a clean-up to ensure proper disposal. It should be noted that in disposal activities, the liability tends to have a "long tail" as to long

---

[2]Id.

term ramifications of such disposal. Incidents such as "Love Canal" and the clean-up efforts currently being undertaken at the Hanford Nuclear Reservation exemplify the long term ramifications.

EMERGENCY MEDICAL SERVICES

The general rule in the area of life saving activities is that a duty to exercise reasonable care under the circumstances is required, whether the individual possesses specialized training or has no training at all. Emergency Medical Service personnel, because of their professional training and because of their position in responding to medical emergencies, are often held to the higher standard than of a reasonable person under the circumstances with the same or similar skills and certifications in the area. (See Professional Standard in Chapter III). For Emergency Medical Technicians (EMT), this means that the EMT is obligated to exercise the same degree of care, skill, and judgment as a comparable first aid provider with the same skills, experience, and certification *as well as* working under the same conditions and circumstances. In some jurisdictions, the EMT is provided specific statutory protections under the doctrine of immunity when the EMT is functioning within the scope of duty. Additionally, fire service organizations often indemnify the EMT against personal liability if the EMT is functioning within the scope of duty. When the EMT ventures beyond these protections or performs activities that are outside the scope of duty, personal liability may take effect against the EMT.

In actions against an EMT or fire service organization for injuries caused by the acts or omission of the EMT, it is important to note that a negligent act by the EMT must cause, or aggravate, the injured individual's injuries in order for liability to attach to the situation. Where the negligent act or omission causes no discernible injury, there is usually no liability. [3]

---

[3]See, for example, *Miller v. Mountain Valley Ambulance Service*, 694 P.2d 364 (Colo. App. 1984).

Failure to respond to an emergency when a duty exists is increasingly becoming the basis for EMT liability. As a general rule, absent statutory immunity, where a special relationship is found to exist between the injured individual and the fire service organization or EMT response group, failure to respond to an emergency medical call can be considered negligence.[4]

As a general rule, the EMT has a duty to respond to and to perform rescue efforts in which he/she is trained and to do so in the specific manner in which his/her training proscribed. Virtually all fire fighters and EMTs are trained in standard first aid and cardiopulmonary resuscitation (CPR). In situations where first aid or CPR may be required and an EMT or fire fighter fails to at least attempt these procedures, this may constitute negligence. If the EMT or fire fighter was not provided with the appropriate equipment to respond safely, such as a one-way airway, the fire service organization may be considered negligent (see Bloodborne Pathogen standard below). Negligence can also happen in situations where CPR or first aid procedures are performed improperly and injury results.

The area of transportation of an injured individual offers the greatest risk for most fire service organizations. Failure to transport, transport to the wrong location and vehicular accident during transport are but a few of the potential risks.[5] Additionally, as EMTs are being provided additional expertise and the technology is evolving to permit additional medical techniques to be performed in the field or during transport, the potential of error, and thus liability, also increases.

EMTs who perform rescue or medical efforts outside the scope of their responsibilities or perform such efforts in a reckless or willful manner, the EMT is usually not protected under the available immunity statute, if any, and is usually not indemnified by the fire service organization. Thus, if the EMT exceeds the proscribed authority or performs activities that are willful or reckless in

---

[4]See, for example, *Archie v. City of Racine*, 627 F.Supp 766 (E.D. Wis. 1986).
[5]See, for example, *Taplin v. Town of Chatham*, 390 Mass. 1, 453 N.E.2d (Mass. 1983).

nature, the EMT is assuming personal liability for his/her own actions.

Fire service organizations should be aware of a new OSHA standard developed to protect fire fighters, EMTs and others from possible exposure to HIV, hepatitis, and other bloodborne diseases. This standard, titled the Bloodborne Pathogen Standard, [6] is especially applicable to EMTs and other first aid responders. The usual areas of exposure include accident sites where medical attention is needed and is provided by an EMT and/or fire service personnel, and other similar occurrences when employees could be exposed to bloodborne pathogens. Additionally, this standard requires the proper labeling and disposal of all medical waste that includes a wide range of items from a blood sample to a used band aid. Below is a synopsis of the procedures acquired to achieve compliance with this new standard:

1. Acquire a copy of this new OSHA standard.
2. Review this standard with your management team and acquire management commitment and appropriate funding for this program.
3. Develop a written program incorporating all required elements of the standard that include, but are not limited to, universal precautions, engineering and work practice controls, personal protective equipment, housekeeping, infectious waste disposal, laundry procedures, training requirements, hepatitis B vaccinations, information to be provided to the physician, medical record keeping, signs and labels, and availability of medical records.
4. Under the area of universal precautions, operations should be analyzed to provide all necessary safeguards to employees who may have possible contact with human blood. Engineering controls and work practices should be examined and evaluated. Procedures should be established for the safe disposal

---

[6]29 C.F.R. §

of used needles, used personal protective equipment, and other equipment. Such areas as the refrigerator, cabinets, or freezers where blood could be stored should be prohibited for the storage of food or drink. Fire fighters and EMTs should be properly trained in the requirements of this standard. Fire fighters and EMTs must be prohibited from eating, drinking, smoking, applying cosmetics or lip balm, or handling contact lenses after possible exposure.

5. Where there is the potential of exposure, personal protective equipment such as surgical gloves, gowns, fluid proof aprons, face shields, pocket masks, ventilation devices, must be provided to employees. This requirement is especially important for your first aid responders, EMTs and other personnel who may be exposed when providing first aid to an injured worker. The personal protective equipment must be appropriately located for easy accessibility. Hypo-allergenic personal protective equipment should be made available to employees who may be allergic to the normal personal protective equipment.

6. Fire service organizations may be required to maintain a clean and sanitary worksite. A *written* schedule for cleaning and sanitizing (disinfected) all applicable work areas must be implemented and included in your program. All areas and equipment that have been exposed must be cleaned and disinfected.

7. All infectious waste, such as bandages, towels, or other items exposed to human blood, must be placed in a closable, leak-proof container or bag with the appropriate label. This waste must be properly disposed of as medical waste according to federal, state, and local regulations. Check with your local hospital or local OSHA office to acquire the regulations applicable to your operations.

8. Contaminated uniforms, smocks, and other items of personal clothing must be laundered in accordance with the standard. *Do not send contaminated uniforms, smocks, or similar articles to your in-plant laundry.*

9. Appropriate EMTs and other fire service personnel should be trained in the requirements of this standard. Where required, appropriate employees who may be exposed are required to undergo a medical examination and acquire a hepatitis B vaccination. Training requirements should include a copy of the OSHA standard and explanation; general explanation of the symptoms and epidemiology of bloodborne diseases; an explanation of the transmission of bloodborne pathogens; an explanation of the appropriate methods for recognizing jobs and other activities involving exposure to blood; an explanation of the use and limitations of practices that will prevent or reduce exposure such as engineering controls and personal protective equipment; Information on the types, proper use, location, removal, handling, decontamination and/or disposal of personal protective equipment; an explanation of the basis for the selection of the personal protective equipment; information on the hepatitis B vaccine, including information on the efficiency, safety, and benefits; information on the appropriate actions to be taken and persons to contact in an emergency; an explanation of the procedure to be followed if an exposure incident occurs, including the method of reporting, the medical follow-up, and medical counseling; and an explanation of the signs and labels. Additional training may be required in applicable laboratory situation and other circumstances.

10. The fire service organization may be required to maintain complete and accurate medical records. These records must include the names and social security numbers of the employees; copy of the employee's hepatitis B vaccination and medical evaluation; copy of the results of all physical examinations, medical testing, and follow-up procedures; the physician's written opinion; and copies of all information provided to the physician. The employer is charged with the responsibility of maintaining the confidentiality of these records.

11. The fire service organization may additionally be required to maintain all training records and these records must include:

dates of all training; names of persons conducting the training; and names of all participants. The training records must be maintained for a minimum of five (5) years. Employees and OSHA must be provided the ability to view and copy these records upon request.

12. All containers containing infectious waste, including but not limited to refrigerators and freezers containing infectious waste, medical disposal containers, and all other containers, *must* be properly labeled. The required label must be fluorescent orange or orange-red with the appropriate symbol and the lettering reading "BIOHAZARD" in a contrasting color. This label *must* be affixed to all containers containing infectious waste and must remain affixed until the waste is properly disposed.

The importance of this OSHA standard is that fire service organizations may now have a duty to comply with this law and failure to comply may provide ample proof in an action for willful or wanton negligence if a fire fighter should acquired HIV or HBV on the job.

DUTY OWED BY OTHERS TO FIRE SERVICE ORGANIZATION

In a majority of the states, fire fighters are considered licensees thus the land owner owes a duty to act as a "reasonable person" in protecting the fire fighter from potential harm while on the property. The land owner must exercise due care and must advise the fire fighter as to the dangers associated with the property or related activities. A small minority of states consider a fire fighter's entry on to property as a special class of entrant and thus entry on to the property is independent of the property owner's permission. Other states have held that since the land owner contacted the fire department, the fire fighters who entered to fight the fire should be considered "invitees." This designation establishes the duty that a land owner owes to the fire fighter when he or she enters privately owned property in an emergency situation.

Under the majority rule, a fire fighter is considered a licensee and thus the fire fighter assumes the basic risks that may be encountered in fighting a fire except where the land owner willfully or wantonly placed the fire fighter in danger of harm. For example, if a fire fighter entered a location where the land owner stored highly explosive chemicals and the land owner failed to inform the fire department of the existence of these chemicals, when the fire fighter is injured because of an explosion of these chemicals, the land owner may be liable to the fire fighter.[7]

Fire fighting is a hazardous occupation where risks are accepted as part of the job. Fire fighters are expected to know the basic risks and common hazards that accompany this occupation.

Land owners usually cannot be held liable for any act or omission that are inherent to the actual fire or emergency and/or is part of the response to the fire. In general, a fire fighter who knowingly and voluntarily has confronted a hazard or risk cannot recover for injuries sustained therefrom.[8] These common hazards and basic risks, such as smoke inhalation and roof collapse, usually do not permit the fire fighter to seek redress from the land owner.[9] However, when a land owner willfully and wantonly exposes a fire fighter to risks that exceed the basic assumed risks and the fire fighter incurs injuries or is killed, liability may rest with the land owner.

In states that consider a fire fighter an invitee, the land owner owes a duty to "exercise reasonable care to keep the premises in safe condition." In the precedent setting case of *Dini v. Naiditch*,[10] a stairway collapses under two fire fighters during a fire. One of the

---

[7]See, *Koehn v. Devereaux*, 495 N.E.2d 211 (Ind. App. 1986)("A fireman is a licensee by permission of law when entering upon a property of another in performance of his duties. Therefore, the fireman may recover only for a positive wrongful act that results in injury").

[8]*Walters v. Sloan*, 20 Cal.3d 199, 571 P.2d 609 (1977).

[9]See, for example, *Lunt v. Post Printing*, 48 Colo. 316, 110 P. 203 (1910).

[10]20 Ill.2d 406, 170 N.E.2d 881 (1960.

fire fighters was killed and the other sustained injuries. The Illinois Supreme Court, in overturning the licensee rule and providing damages to the family of the deceased fire fighter and the injured fire fighter stated, "It is our opinion that since the common law rule labeling fireman as licensee is but an illogical anachronism, originating in a vastly different social order, and pock-marked by judicial refinements, it should not be perpetuated in the name of stare decisis. That doctrine does not confine our courts to the 'calf path' nor to any rule currently enjoying a numerical superiority of adherents. Stare decisis ought not to be the excuse for a decision where reason is lacking."[11]

This decision by the Illinois Supreme Court was based on the fact that the land owner failed to provide the appropriate fire doors, did not have fire extinguisher, permitted trash to accumulate, and stored flammable liquids below the wooden stairs that ultimately collapsed.

Under a doctrine known as the "Fireperson's Rule" (part of the Rescue Doctrine, also known as the "Fireman's Rule), the person committing the harm cannot be held responsible for any injuries suffered by fire fighters or other public safety personnel if his/her only act of wrongdoing was the direct and proximate cause of the emergency that necessitated the fire department (or other public safety organization) response. In essence, the fire fighter assumes a certain amount of the risks that are inherent in the job and thus if the land owner created no other or additional risks, the land owner cannot be liable for injuries incurred by the fire fighter in the course of performing his or her job.[12]

Where the actions of the land owner are willful and wanton, the majority of the courts have permitted recovery. However, some courts have precluded recovery for intentional acts, such as arson, and often the distinction between whether the act is willful or sim-

---

[11]Id.

[12]See, *Armstrong v. Mailand*, 284 N.W.2d 343 (Minn. 1979); *Moreno v. Marrs*, 102 N.M. 373, 695 P.2d 1322 (N.M. App. 1984).

ply negligent is the determining factor. The "fine line" between a willful act and a negligent act is often dependant upon the facts of the situation.

The application of the Fireperson's Rule serves to basically immunize the landowner from liability. Thus careful analysis of the situation, applicable statutes, and case law is necessary in order to determine the applicability of this Rule. Any analysis of this situation should include the following:

1. Does the jurisdiction use the Fireperson Rule?
2. Did the landowner fail to warn the fire fighter of hidden dangers?
3. Did the landowner know or should have known of the hidden dangers?
4. Did the landowner warn the fire service organization?
5. Did the landowner affirmatively misrepresent the dangers?
6. Did the land owner falsify causes of the injuries?
7. Did the landowner falsify documents regarding the dangers?
8. Were the fire fighter's injuries a direct result of the dangers?
9. Did the landowner maintain the entire property in a safe condition, including fire escapes, sprinkler systems?
10. Did the landowner deny access to the fire fighter any part of the property?
11. Did the landowner interfere with the fire service organization's performance?
12. What is the legal status of the fire fighter in the jurisdiction?

Many states have modified or abolished the Fireperson's Rule thus a careful analysis of the state statutes and case law is essential in determining the applicability of this Rule to the situation. The trend among the courts and legislatures appears to be that the Fireperson's Rule is extremely harsh in barring all recovery by the injured fire fighter.

When the property owner is the municipality, such as a fire house, the land owner owes a duty of care with respect to the maintenance, care, and conditions of the property. Thus, in most public sector fire service organizations, the fire service organization owes a reasonable duty of care to the fire fighters to maintain safe facilities and equipment. For utilities, such as electric companies, natural gas companies, the utility owes a duty to the fire fighter to keep all dangerous equipment and other property in reasonably safe condition. Usually, inspections for potential hazards are required by industry practice or statute. Statutory or regulatory requirements usually require the utility to respond to dangerous conditions in a reasonable time.[13]

SELECTED CASES

(This case has been edited for the purposes of this text)

Seibert Security Services, Inc., Petitioner
v.
Superior Court of the state of California
for the County of San Bernardino, Respondent
John **Migailo** et al., Real Parties in Interest
Court of Appeal, Fourth District, Division 2
August 3, 1993
Dabney, Acting Presiding Justice

Petitioner Seibert Security Services, Inc. (hereinafter "Seibert"), a defendant in the action below, seeks reversal of an order denying its motion for summary judgment and/or adjudication of issues. Seibert's contention is that it established, as a matter of law, a complete defense in that plaintiff and real party in interest John Migailo's action is barred by the so-called "fireman's rule."

---

[13]See, for example, *Verges v. New Orleans Public Service*, 455 So.2d 878 (La.App. 1984)(Utility acted reasonably when boom came in contact with overhead power line causing injury); *Pennebacker v. San Joaquin Light and Power*, 158 Cal. 579, 112 P. 459 (1910)(Fire fighter electrocuted by wire falling in fire—no negligence by utility).

Facts

At the time of his injury, Migailo was a police officer employed by the City of San Bernardino. He had brought an arrested suspect to the San Bernardino County Hospital for examination of possible injuries. While there, a mental patient, defendant Raymond Shyptycki, became abusive towards a black security officer employed by Seibert (Grigsby) and a black police officer. At the time, Shyptycki was restrained in a chair; Migailo asked that he be handcuffed for greater control, but Grigsby failed to do so. Shortly thereafter, Shyptycki stood up and grabbed for Grigsby's baton, and Migailo helped subdue him.

Shyptycki was then handcuffed and put in an isolation cell, but the handcuffs were taken off because he seemed "pretty pleasant" to another Seibert employee, Timothy Leggett. Within fifteen minutes, Shyptycki attacked Leggett, who called for help. Officer Migailo responded and was injured while subduing Shyptycki. During the struggle between Shyptycki and Migailo, Leggett dropped back and did not assist Migailo, although another Seibert employee did participate.

The trial court denied Seibert's motion on two grounds: (1) that triable issues remained as to whether Migailo's presence was "independent and unrelated to the conduct that proximately caused plaintiff's injuries"; and (2) as to whether the conduct "proximately causing the injury occurred after the defendant ... knew or should have known of the presence of the plaintiff, a peace officer ... "

Discussion

The "fireman's rule" was born almost a century ago, earning nearly unanimous acceptance (see *Walters v. Sloan,* (1977) 571 P.2d 609). Although this rule has been attacked in recent years, the *Walters* court solidly reaffirmed its continuing viability in this jurisdiction. The rule applies equally to police officers injured in the course of their duties. Although the rule was originally framed with reference to a landowner's premises liability, the rule is fundamentally based on public policy and the nature of the relationship between firefighter or police officer and the public. The rule is designed with the recognition that most fires are due at least in part to negligence, " . . . and in the final analysis the policy decision is that it would be too burdensome to charge all who carelessly cause or fail to prevent fires with the injuries suffered by the expert retained with public funds to deal with those inevitable, although negligently created occurrences." The freedom from liability provided by the fireman's rule "permits individuals who require police or fire department assistance to summon aid without pausing to consider whether they will be held liable for consequences that, in most cases, are beyond their control."

A number of exceptions to the fireman's rule exist such as those contained in Civil Code section 1714.9. That statute abrogates the fireman's rule if "the conduct causing the injury occurs after the person knows or should have known of the presence of a peace officer, firefighter, or emergency medical personnel."

The second express basis for the trial court's ruling was that Migailo's presence was unrelated to the negligence that caused his injury. This exception applies when, for example, a police officer, pursuing a suspect, is injured because of a dangerous condition on real property. In the same vein is the case of a firefighter who, while fighting a fire, falls through a defectively maintained roof. The rationale of these cases is that, as to the suffering of such injuries, the firefighter or police officer stands in the same posture as any other citizen venturing onto the land of another; his occupation compels him to face felons or fires, not rickety roofs or faulty fences.

The law now warns any person causing an incident that results in the summoning of a police officer, firefighter or emergency medical care provider—or any person present when such a person arrives to render services—that tort liability may be imposed for wrongful acts committed in the knowledge that the person covered by the statute is present. For example, a suspect or prisoner who resists a recognized officer (or any officer the suspect *should* recognize as an officer) is liable for any injury caused by his resistance.

Read literally, the statute appears to except the situation represented here from the operation of the fireman's rule. It is not disputed that Seibert personnel were aware of Migailo's presence. However, in our view such a construction leads to absurd results and is contrary to any reasonable policy of the law.

Under the fireman's rule, a person's negligence cannot be used to impose tort liability if the officer or firefighter is injured in remedying the condition negligently caused. However, under the construction urged by the real parties, every person would be under a new, special duty of care whenever the police officer or firefighter is "present" and the person knew or should have known of the presence. While such an exception can reasonably be applied to *new* conduct committed by such person after the officer has been summoned in the course of his duties, it *cannot* be applied to conduct occurring while the officer is "present" but before he has undertaken his duties of protection or crime prevention. For example, A driver on a mountain road idly observes a fire truck (obviously containing firefighters, but not proceeding to a call) on the road immediately behind him. The driver carelessly flicks a lighted cigarette butt out the window causing a brush fire. The fire truck immediately stops and the firefighters fight the fire; one is injured. Is the driver liable?

In this hypothetical case, the defendant was aware that a member of the protected class was present, and nevertheless committed negligent conduct that raised some risk that intervention would be necessary. Strictly speaking, Civil Code section 1714.9, subdivision (a) (1) applies. However, this approach is fraught with inequities.

It is unequitable as to the defendant, because the same conduct, committed with the same likelihood that *some* firefighter would have to be summoned, would be immune from liability if no specific member of the protected class was present, or even if no such person was present, but was not reasonably known to be so . . . (e.g., a plainclothes police officer). As long as conduct is merely negligent, and not wilful or malicious, there is no point in imposing liability on the defendant whose acts are likely to involve a specific police officer, while precluding liability on the defendant whose negligent acts require the summoning of an officer at random.

In our view, the statute (Civil Code section 1714.9) applies only to conduct committed after the officer responds to a call for assistance, or while he is in the performance of his duties with respect to a specific incident, and such conduct increases the risk of injury to the officer. It does not require citizens to be especially vigilant or careful whenever they happen to observe the near presence of a member of the class covered by the statute.

In this case, Migailo was performing one duty—completing paperwork relating to the injured suspect—when the alleged negligence of Seibert caused him to initiate a new and different law enforcement action and attempt to subdue Shyptycki. While the conduct of Seibert may have been "independent of and unrelated to" the conduct that originally brought plaintiff to the hospital, it is factually undisputed that it was the immediate cause of Migailo's presence in or near the holding cell of Shyptycki. The chance presence of such personnel cannot mean that any negligent conduct that creates a crisis to which such personnel react becomes actionable in tort. Unless a police officer or firefighter has come to a specific location to perform a specific immediate duty, and the defendant's unrelated negligent or intentional conduct increases the risks inherent in performing that duty, this exception is inapplicable.

Disposition

Let a peremptory writ of mandate issue, directing the superior court to vacate its order denying petitioner's motion for summary judgment, and to enter a new order granting said motion.

(This case has been edited for the purposes of this text.)

McCoy
vs.
The State
262 Ga. 699

Appellant was indicted for felony murder and arson in the first degree. His first trial resulted in a conviction for arson in the first degree and a mistrial on the felony murder charge because of the jury's inability to reach a verdict. A second trial on the felony murder count resulted in a verdict of guilty. The trial court merged the arson conviction into the felony murder conviction and sentenced only for the felony murder. Appellant asserts on appeal that the trial court erred in refusing to give certain requested jury instructions and that the evidence did not support the verdict.

The evidence at trial authorized the jury to find that the appellant and his co-indictee left a party and walked to an abandoned house. After exploring the house, and having noticed the presence of a well closed by a wooden cover, appellant borrowed his companion's lighter and deliberately set the house afire. Two volunteer firemen who responded to the fire were directed to take a hose to the back of the house to prevent the fire from spreading to other property. In the darkness and the dense smoke from the fire, one of the firemen fell into the well, the cover of which had been burned in the fire. The well was filled with smoke and ashes and the fireman, unable to obtain sufficient oxygen, died of acute carbon monoxide poisoning associated with smoke inhalation and oxygen depletion.

It being clear from the evidence that appellant deliberately set the house afire, that the victim came to the scene as a direct result of appellant having set the fire, that the protective cover over the well was burned away by the fire appellant set, and that the victim died as a result of breathing the concentrated smoke from the fire appellant set, we hold that the evidence was sufficient to authorize a rational trier of fact to find appellant guilty beyond a reasonable doubt of felony murder with arson in the first degree as the underlying felony. Appellant's reliance on *State v. Crane*, 279 S.E. 2d 695 (1981), and *Hill v. State*, 295 S.E. 2d 518 (1982), is unwarranted. The felony murder statute was inapplicable in those cases because the deaths were not caused by the defendant but by the victim and a police officer, respectively, whereas the death in this case was directly attributable to appellant's felonious conduct in setting the fire.

In support of his effort to show that the victim's death was not due to his criminal conduct in burning the house, but to the negligence of the landowner in leaving an abandoned well unfilled, appellant requested a charge on the duty to fill in abandoned wells. The requested charge, however, was not a complete and accurate statement of the law in that it did not specify whose responsibility it was to fill in abandoned wells; indeed, it could be inferred from the requested charge and the evidence that appellant, having discovered the well, was under a duty to report it. Since the requested charge was not accurate and was not adjusted to the evidence, there was no error in refusing to give it.

Appellant requested charges on the offenses of involuntary manslaughter, reckless conduct, and criminal trespass, arguing that his conduct could have been found to constitute one of the latter two misdemeanors, authorizing the jury to find him guilty of involuntary manslaughter rather than felony murder. However, the uncontradicted evidence in this case showed the completion of the greater offense of arson in the first degree, rendering it unnecessary that the trial court charge on the lesser offenses.

Judgment affirmed.

9

Liabilities in Hiring

It usually takes 100 years to make a law, and then, after it's done its
work, it usually takes 100 years to be rid of it.

Henry Ward Beecher

Potential liabilities in the areas of hiring, screening, promotion and
other related aspects of the employment relationship have
increased significantly for fire service organizations in recent
years. Congress and states have provided specific rights to individuals and/or expanded existing rights that vest at different times in
the employment relationship ranging from the time of advertise-
ment of the position through the termination of the employment
relationship. In this balance between employers' rights and
employees' rights in the workplace, the current trend reflects a significant shift increasing the rights of the individual in the workplace.

The reasons for this escalation of individual rights in the workplace appear to be the reaction of government to a number of factors including the decline of protections provided under a collective bargaining (union) relationship, a shifting of workplace
demographics, reliance upon employers for health care benefits,
reliance on employers for retirement benefits, and other related factors. Fire service organizations should be cognizant of these shifting winds and make the necessary modifications to ensure compliance with the various laws within this area.

WRONGFUL HIRING

The employment area can be a mine field for the unknowing fire service organizations. One of the newest and novel theories to evolve in recent years has been that of wrongful hiring. Under this theory, an organization may be liable for hiring an individual without properly screening the individual's background. In most cases, the organization assumes liability when the individual who is hired for the position has some type of criminal or related background that may place other employees, other individuals, or the public at risk. When the organization is found to be negligent in failing to detect this propensity of the employee in the background screening, potential liability may be found against the organization. In *Ponticas v. K.M.S. Investments,*[1] 331 N.W. 2d. 907 (Minn. 1983), an employer was held liable to the victims under a negligence theory for hiring a dangerous employee with a criminal record as an apartment manager and the employee later raped a resident of the apartment complex. Under this theory, fire service organizations may has a duty to make at least a cursory evaluation of the individuals who they hire to ensure that the individual is not posing a risk to other employees or the general public. The new theory has now placed fire service organizations and all employers in a precarious position. If the fire service organization does not perform any type of background screening of prospective fire fighters, the potential for a wrongful hiring is present. Conversely, as can be seen later in this chapter, if a fire service organization delves too far into the background of fire fighter, the fire service organization may be liable in other areas for personal invasion of privacy. Fire service organizations must take extreme caution in evaluating new employees for employment in their organization and promotion into higher level positions. Fire service organizations are now "walking the tight rope" of potential liability on several fronts in the employment arena. Fire service organizations must know the

---

[1]331 N.W. 2d 907 (Minn. 1983).

rules of the game in order to properly and legally circumnavigate this evolving area of employment law.

CIVIL RIGHTS VIOLATIONS

The enactment of the Civil Rights Act of 1964[2] propelled the area of workplace discrimination to the forefront. The Civil Rights Act of 1964 barred discrimination in virtually all settings based upon race, color, religion, national origin and sex. Of particular importance in the employment setting is Title VII of the Civil Rights Act of 1964.[3] Title VII bars all types of discrimination based on any of the protected classes (race, religion, color, national origin or religion), in all fire service organizations. Thus, fire service organizations are required to create a non-discriminatory work environment.

The purpose of Title VII of the Civil Rights Act of 1964 is to remove the artificial, arbitrary, and unnecessary barriers to employment when such impediments operate invidiously to discriminate against individuals on the basis of racial or other impermissible classifications.[4]

Title VII of the Civil Rights Act of 1964 was amended in 1972 by the Equal Employment Opportunity Act[5] and most recently by the Civil Rights Act of 1991. The Civil Rights Act of 1991 provides significant changes to Title VII of the Civil Rights Act of 1964 including the following:

1. Recovery of compensatory and punitive damages is now permitted and claims can be tried before a jury. Damages are capped on a sliding scale dependant on the size of the employer and damages are expressly unavailable in disparate impact litigation.

---

[2] 42 U.S.C.A. §2000e-2
[3] Id.
[4] *Griggs v. Duke Power Co.*, 401 U.S. 424 (1971).
[5] P.L. 92-261 (Mar. 24, 1972) 86 Stat. 103.

2. The meaning of "business necessity" and other burden of proof rules in disparate impact litigation in most instances are restored to their pre *Wards Cove* status.

3. The adjustment of test scores or the use of different cut-off scores on employment related tests on the basis of race, sex, color, religion or national origin is prohibited.

4. Employment actions may now violate the Civil Rights Act even though the same action would have been taken absent discriminatory motive. In these cases, known as "mixed motive" cases, remedies are limited to injunctive relief, declaratory relief and attorney's fees.

5. Extraterritorial application of Title VII has been expanded and clarified.

6. Expert witness fees can not be awarded in addition to attorney fees.

7. The time limitation for filing suits against the federal government when the federal government is the employer has been extended to ninety days.

Note: The Civil Rights Act of 1991 also directly and indirectly affects the Americans With Disabilities Act, Rehabilitation Act of 1973, Age Discrimination in Employment Act of 1967, and other laws.

The agency charged with the enforcement of the Civil Rights Act is the Equal Employment Opportunity Commission (also known as the EEOC). The EEOC is charged with investigating any complaints by individuals of a protected class who believe that they have been discriminated against in the workplace. The EEOC is empowered to act and to investigate employers, records, and even go so far as to represent the employee against the employer. It should be noted that the EEOC, like many government agencies, promotes the use of alternate dispute mechanisms to settle such conflicts between employers and employees.

A fire service organization falls within the jurisdiction of Title VII if:

- The organization employs 15 or more employees;
- Employees have been employed for each working day in each of 20 or more calendar weeks in the current or preceding calendar year;
- The organization engages in an industry affecting commerce;[6]
- Employer includes employment agencies, labor organizations, and employees of state and local governments, governmental agencies, political subdivisions, and other related divisions with certain exceptions.

Title VII defines the following as unlawful employment practices for an employer to:

- fail or refuse to hire or to discharge any individual, or otherwise to discriminate against any individual with respect to his or her compensation, terms, conditions, or privileges of employment, because of such individual's race, color, religion, sex, or national origin
- limit, segregate, or classify his or her employees or applicants for employment in any way that would deprive or tend to deprive any individual of employment opportunities or otherwise adversely affect his or her status as an employee, because of such individual's race, color, religion, sex or national origin[7]

The EEOC usually distinguishes employer policies that are discriminatory "on their face" (known as disparate treatment) from rules and policies that are facially neutral but nonetheless have a disproportionate impact on protected classes (known as disparate impact). Fire service organizations that enforce policies and rules that adversely impact or directly discriminate against protected class fire fighters may be violating Title VII.

---

[6]42 U.S.C.A. § 2000e(b).

[7]42 U.S.C.A. § 2000e-2(a).

Of particular importance to fire service organizations is the job-relatedness standard within Title VII. Title VII prohibits job requirements that have a disparate impact on individuals as a result of their protected characteristics under Title VII, unless the fire service organization can prove that the job requirement is job-related. Below are some of the job requirements that have been the source of EEOC litigation:

- Education—As a general rule, the extent to which an employer may use education as a job requirement varies with the public's interest of health and safety in the performance of the job.[8] Given the expansive public interest in health and safety in the fire service, this requirement is normally considered job related.
- Health Requirements—A fire service organization is permitted to reject a prospective fire fighter who fails the physical examination as long as the physical impairment or disability would prevent the applicant from performing the basic job functions. Fire service organizations must provide reasonable accommodations and meet the other requirements of the ADA (See Section on the Americans With Disabilities Act).
- Strength Requirements—A fire service organization may reject an applicant who fails a strength test if strength is a bona fide, job-related qualification (used for measurement rather than in the abstract).[9]
- Height and Weight Requirements—Height and weight specifications that deny equal employment opportunities to all groups of individuals who are protected under Title VII are unlawful unless such qualifications are necessary to perform the job in question.[10]
- Work Experience—A fire service organization may require

---

[8]See, *Townsend v. Nassau County Medical Center*, 558 F.2d 117 (CA-2, 1977).

[9]*Dothaard v. Rawlinson*, 433 U.S. 321 (1977).

[10]*Davis v. County of Los Angeles*, 566 F.2d 1334 (CA-9, 1977)

previous work experience as a valid and relevant job requirement so long as this requirement relates to the successful performance of the job in question [11] or is recognized as a business necessity.

* Appearance and Dress Requirements—As a general rule, fire
service organizations cannot reject an applicant because his or
her appearance is typical of minority candidates. However, fire
service grooming regulations usually do not violate Title VII
even though the proscribed dress standards may differ somewhat for male and female as long as the dress requirements are
reasonably related to the fire service organization's needs and
are commonly accepted norms in the fire service.[12]

* Hair Requirements—Given the nature of the activity in the fire
service, the length of hair and facial hair issues are normally
addressed within the realm of personal safety. (For example,
beards are normally prohibited because a fire fighter cannot
achieve an adequate seal on the self-contained breathing apparatus. As a general rule, the EEOC generally considers it a violation of Title VII whenever an employer's grooming code
requires men to have different hair styles than women.[13]

* Credit Requirements—Requiring that an applicant have a
good credit record IS usually unlawful unless business necessity is proven.

* Arrest and Criminal Records—Because of the type of work
performed by fire service organizations, a "clean" arrest record
is normally required for safety and business necessity reasons.
Because of the disproportionate impact on black candidates,
the EEOC considers a requirement that applicants have no previous arrest record as an unlawful job qualification unless the
employer can show the necessity in the operations of the particular business. Additionally, the refusal to hire an applicant

---

[11]*Griggs v. Duke Power,* supra.

[12]See, *Caroll v. Talman Fed. Sav. & Loan Assn.,* 604 F.2d 1028 (CA-7, 1979).

[13]See, EEOC Decision No. 72-2179 (1972).

on the basis of a conviction of a crime may be unlawful unless the employer is able to show that he or she has first considered the circumstances surrounding the particular case and employment of the individual would be inconsistent with the safe and efficient operation of the business.[14]

- Military Record—A fire service organization that disqualifies an prospective fire fighter because he or she has received a less-than-honorable discharge from the military may be guilty of unlawful discrimination unless the fire service organization can prove that the rejection is related to job performance.[15]

- Alienage—Citizenship requirements are unlawful whenever they have a purpose or effect of discriminating against an individual on the basis of national origin.

- Language Requirements—Proficiency in English as a job requirement may be unlawful unless business necessity or job relatedness is shown.

- No Spouse Requirements—Generally, fire service policies prohibiting the hiring of a spouse of a fire fighter is lawful so long as the rule is neutral as to sex.

- Marriage Status—Policies that forbid the hiring of married women but do not forbid the hiring of married men are usually unlawful.

- Pregnancy—With the amendment to Title VII in 1978, employers may not discriminate against women who are affected by pregnancy, childbirth, or related medical conditions. Pregnancy discrimination is encompassed under the sex prohibition under Title VII and many states have adopted pregnancy discrimination statutes. (Also see Family and Medical Leave Act).

- Sex Status—It is unlawful for a fire service organization to hire a certain number of male fire fighters and a certain number of female fire fighters unless the fire service organization

---

[14]EEOC Decision No. 78-10 (1977).

[15]*Dozier v. Chupka*, 395 F.Supp. 836 (DC Ohio 1975).

can show that this requirement is a bona fide occupational qualification.

- Religious Convictions—A fire service organization must make reasonable accommodations to the religious needs of employees and may not reject candidates because of their religious needs unless such accommodations would create an undue hardship on the organization.
- Age Requirements—Employment discrimination on account of age between 40 and 70 violates the Age Discrimination in Employment Act, age restrictions may also violate Title VII where the restrictions apply only to employees within a protected class.

In testing or other selection procedures for new candidates, if the test adversely affects a disproportionate number of persons protected under Title VII, it may be an unlawful selection method unless the test is related to the successful performance of the job or the test is used is necessary to the operations of the fire service organization. The EEOC has accepted three methods of validating selection tests:

- criterion-related validations—compares criteria with successful job performance
- construct validation—compares mental and psychological traits
- content validation—test closely duplicates job performance

Fire service organizations should be aware that Title VII protections extend to work assignments, promotions, transfers, discipline and other areas of the employment setting. Fire service organizations must maintain a work environment free of racial, sex, religious or ethnic harassment. The fire service organization is under a duty to take reasonable measures to attempt to control or eliminate overt expressions of harassment in the workplace. For example, a fire service organization may have a duty to remove sexually

explicit photographs, calendars, and magazines (i.e., *Playboy* centerfolds depicting females or *Playgirl* centerfolds depicting males)
from the fire house walls or lockers and establish policies to prevent future posting. (See Discrimination in Chapter IX for EEOC
claims procedures).

Although Title VII is the most visible are of potential discrimination in the fire service, other potential civil rights liabilities also
exist for fire service organizations. A common federal civil rights
action against governmental agencies arises under 42 U.S.C. §
1983 and involves a claim that an individual was deprived of a constitutional right by an individual or organization acting under the
color of law. These actions are often based on the same conduct
that forms the basis for a tort action under a state law. A § 1983
action can be brought for misuse of legal authority, even in violation of state law, and there is no general requirement that a state
bring the initial action.

This statute provides that

> every person who, under the color of any statute, ordinance, regu-
> lation, custom, or usage, of any State or Territory, subjects, or causes
> to be subjected, any citizen of the United States or other person
> within the jurisdiction thereof to the deprivation of any rights, privi-
> leges, or immunities secured by the Constitution and laws, shall be
> liable to the party injured in an action at law, suit in equity, or other
> proper proceeding for redress.[16]

Section 1983 creates no rights for the individual in and of itself
but is a vehicle to redress violations of Constitutional rights and
certain federal statutes. Section 1983 actions have been prevalent
to date against police departments and various other governmental
agencies for such actions as illegal searches, use of force issues,
and electronic surveillance issues.

Fire service officers should be aware that superior officers may
be liable for participation in unconstitutional conduct, for grossly
negligent failure to train or discipline, or for the promulgation of

---

[16]42 U.S.C. § 1983.

policies that cause constitutional violations. However, officers cannot be liable under § 1983 under the doctrine of *respondeat superior*. Additionally, a superior officer who leads or directs others who commit constitutional violations may be liable even if he or she is not at the scene. [17]

In most circumstances, states and state fire service agencies are not considered proper defendants in § 1983 actions, however state fire officers may be proper in certain circumstances. Local governments and fire services organizations attached to local governments can be liable for § 1983 actions if they adopt or tolerate policies or customs that have caused a wrong. Again, the doctrine of *respondeat superior* is normally not applicable in these types of circumstances.

AMERICANS WITH DISABILITIES ACT

Overview and Impact

With the signing of the *Americans With Disabilities Act of 1990* (known as the "ADA"), discrimination against qualified individuals with physical or mental disabilities in all employment settings was prohibited and additional amendments to the Rehabilitation Act of 1973 were provided. Given the impact of the ADA on the job functions of a fire service organization, especially in the areas of workers' compensation, restricted duty programs, facility modifications, and other areas, it is important for fire service organizations to have a firm grasp of the requirements of this new law.

The ADA is divided into five (5) Titles of which all Titles appear to have the potential of substantially impacting a covered public or private sector fire service organization. Title I contains the employment provisions that protect all individuals with disabilities who are in the United States, regardless of national origin and immigration status. Title II prohibits discrimination against qualified indi-

---

[17]See, *Specht v. Jensen*, 832 F.2d 1516 (10th Cir. 1987) (Superior officer who told others to "take care of it" set in motion a series of constitutional violations).

viduals with disabilities or excluding qualified individuals with disabilities from the services, programs, or activities provided by public entities. Title II includes the transportation provisions. Title III, entitled "Public Accommodations," requires that goods, services, privileges, advantages, or facilities of any public place to be offered "in the most integrated setting appropriate to the needs of the individual."[18] Title III additionally covers transportation offered by private entities. Title IV addresses telecommunications. Title IV requires that telephone companies provide telecommunication relay services and television public service announcements produced or funded with federal money include closed caption. Title V includes the miscellaneous provisions. This Title noted that the *Americans With Disabilities Act* does not limit or invalidate other federal and state laws providing equal or greater protections for the rights of individuals with disabilities and addresses related insurance, alternate dispute, and congressional coverage issues.

Title I of the ADA goes into effect for *all* employers and industries engaged in interstate commerce with 25 or more employees on July 26, 1992 and on July 26, 1994, the ADA will become effective for all employers with 15 or more employees.[19] Just as fire service organizations were not exempt from the ADA's predecessor, the Rehabilitation Act of 1973, most fire service organizations will be subject to the ADA mandates.

Title II that applies to public services[20] and Title III requiring public accommodations and services operated by private entities becomes effective on January 26, 1992,[21] except for specific subsections of Title II that went into effect immediately on July 26, 1990.[22] A telecommunication relay service required by Title IV must be available by July 26, 1993.[23]

---

[18]ADA Section 305

[19]ADA Section 101(5), 108, 42 U.S.C. 12111.

[20]ADA Section 204(a), 42 U.S.C. 12134.

[21]Id.

[22]ADA Section 203 (a), 306 (a), 42 U.S.C. 12186.

[23]ADA Section 102 (a), 42 U.S.C. 12112.

Title I prohibits covered employers from discriminating against a "qualified individual with a disability" with regard to job applications, hiring, advancement, discharge, compensation, training, and other terms, conditions, and privileges of employment.[24]

Section 101 (8) defines a "qualified individual with a disability" as any person

> who, with or without reasonable accommodation, can perform the essential functions of the employment position that such individual holds or desires ... consideration shall be given to the employer's judgment as to what functions of a job are essential, and if an employer has prepared a written description before advertising or interviewing applicants for the job, this description shall be considered evidence of the essential function of the job.[25]

The Equal Employment Opportunity Commission (known as the EEOC) provides additional clarification as to this definition in stating "an individual with a disability who satisfies the requisite skill, experience and educational requirements of the employment position such individual holds or desires, and who, with or without reasonable accommodation, can perform the essential functions of such position."[26]

Congress did not provide a specific list of disabilities covered under the ADA because "of the difficulty of ensuring the comprehensiveness of such a list."[27] Under the ADA, an individual has a disability if one of the following three statements applies to the individual:

1. has a physical or mental impairment that substantially limits one or more of the major life activities of such individual
2. has a record of such an impairment
3. is regarded as having such an impairment[28]

---

[24]Id.

[25]ADA Section 101 (8).

[26]EEOC Interpretive Rules, 56 Fed. Reg. 35 (July 26, 1991).

[27]42 FR 22686 (May 4, 1977); S. Rep. 101-116; H. Rep. 101-485, Part 2, 51.

This definition parallels the language defining an "individual with a handicap" in the Rehabilitation Act.[29] For an individual to be considered "disabled" under the ADA, the physical or mental impairment must limit one or more "major life activities." Under the U.S. Justice Department's regulations issued for Section 504 of the Rehabilitation Act, "major life activities" is defined as, "functions such as caring for one's self, performing manual tasks, walking, seeing, hearing, speaking, breathing, learning and working."[30] Congress clearly intended to have the term "disability" construed broadly. However, this definition includes neither simple physical characteristics, nor limitations based on environmental, cultural, or economic disadvantages.[31]

A substantial issue under the Rehabilitation Act, and one left unclarified under the ADA, is whether a condition is covered when it limits the individual's access to one or more types of employment but otherwise does not limit the individual's employability or otherwise impairs a major life activity. The Rehabilitation Act case most often cited in is *E.E. Black, Limited v. Marshall*.[32] In finding that working was a major life activity, the *Black* court listed the following factors for determining whether a individual's impairment substantially limits employment opportunities:

- the number and types of jobs from which the impaired individual is disqualified
- the geographical area to which the applicant has reasonable access
- the applicant's own job expectations and training
- the criteria or qualifications in use generally
- the types of jobs to which the rejection would apply

[28]SubTitle A, Section 3(2). The ADA departed from the Rehabilitation Act of 1973 and other legislation is using the term "disability" rather than "handicap."

[29]29 U.S.C. Section 706(8) (Supp. V. 1975).

[30]28 C.F.R. Section 41.31. This provision is adopted by and reiterated in the Senate Report at page 22.

[31]See, *Jasany v. US Postal Service*, 755 F2d 1244 (6th Cir. 1985).

[32]497 F. Supp. 1088 (D. Haw. 1980).

The second prong of this definition is "a record of such an impairment disability." The Senate Report and the House Judiciary Committee Report each stated

> This provision is included in the definition in part to protect individuals who have recovered from a physical or mental impairment that previously limited them in a major life activity. (Discrimination on the basis of such a past impairment would be prohibited under this legislation. Frequently occurring examples of the first group, i.e., those who have a history of an impairment, are people with histories of mental or emotional illness, heart disease or cancer; examples of the second group, i.e., those who have been misclassified as having an impairment, are people who have been misclassified as mentally retarded.)[33]

The third prong of the statutory definition of a disability extends coverage to individuals who are "being regarded as having a disability." The ADA has adopted the same "regarded as" test used for Section 504 of the Rehabilitation Act:

> "Is regarded as having an impairment" means (A) has a physical or mental impairment that does not substantially limit major life activities but is treated ... as constituting such a limitation; (B) has a physical or mental impairment that substantially limits major life activities only as a result of the attitudes of others toward such impairment; (C) has none of the impairments defined (in the impairment paragraph of the Department of Justice regulations) but is treated ... as having such an impairment.[34]

Under EEOCs regulations, this third prong covers three classes of individuals:

- Persons who have physical or mental impairments that do not limit a major life activity but who are nevertheless perceived

---

[33]S. Rep. 101-116, 23; H. Rep. 101-485, Part 2, 52-3.

[34]45 C.F.R. 84.3 (j)(2)(iv), quoted from H. Rep. 101-485, Part 3, 29; S. Rep. 101-116, 23:H. Rep. 101-485, Part 2, 53; Also see, School Board of Nassau County, Florida v. Arline, 107 S. Ct. 1123 (1987)(leading case).

by covered entities (employers, places of public accommodation) as having such limitations. (For example, an employee with controlled high blood pressure that is not, in fact, substantially limited, is reassigned to less strenuous work because his employer's unsubstantiated fear that the individual will suffer a heart attack if he continues to perform strenuous work. Such a person would be "regarded" as disabled.)[35]

- Persons who have physical or mental impairments that substantially limit a major life activity only because of a perception that the impairment causes such a limitation. (For example, an employee has a condition that periodically causes an involuntary jerk of the head, but no limitations on his major life activities. If his employer discriminates against him because of the negative reaction of customers, the employer would be regarding him as disabled and acting on the basis of that perceived disability.)[36]

- Persons who do not have a physical or mental impairment but are treated as having a substantially limiting impairment. (For example, a company discharges an employee based on a rumor that the employee is HIV-positive. Even though the rumor is totally false and the employee has no impairment, the company would nevertheless be in violation of the ADA.)[37]

Thus, a "qualified individual with a disability" under the ADA is any individual who can perform the essential or vital functions of a particular job with or without the employer accommodating the particular disability. The fire service organization is provided the opportunity to determine the "essential functions" of the particular job before offering the position through the development of a *Written Job Description*. This written job description will be considered evidence to which functions of the particular job are essential and

---

[35]EEOC Interpretive Guidelines, 56 Fed. Reg. 35, 742 (July 26, 1991).

[36]S. Comm. on Lab. and Hum. Resources Rep. at 24; H. Comm. on Educ. and Lab. Rep. at 53; H. Comm. on Jud. Rep. at 30-31.

[37]29 CFR Section 1630.2(1).

which are peripheral. In deciding the "essential functions" of a particular position, the EEOC will consider the employer's judgment, whether the written job description was developed prior to advertising or beginning the interview process, the amount of time spent on performing the job, the past and current experience of the individual to be hired, relevant collective bargaining agreements and other factors.[38]

The EEOC defines the term "essential function" of a job as meaning "primary job duties that are intrinsic to the employment position the individual holds or desires" and precludes any marginal or peripheral functions that may be incidental to the primary job function."[39] The factors provided by the EEOC in evaluating the "essential functions" of a particular job include the reason the position exists, the number of employees available, and the degree of specialization required to perform the job.[40]

Congress was particularly concerned about the treatment of the disabled individual, who, as a matter of fact or employer prejudice, was believed to be a direct threat to others. To address this issue, Congress provide that any individual who poses a direct threat to the health and safety of others that cannot be eliminated by reasonable accommodation may be disqualified from the particular job.[41] The term "direct threat" to others is defined by the EEOC as meaning "a significant risk of substantial harm to the heath and safety of the individual or others that cannot be eliminated by reasonable accommodation."[42] The determining factors to be considered include the duration of the risk, the nature and severity of the potential harm, and the likelihood the potential harm will occur.[43]

Additionally, the EEOC's Interpretive Guidelines state:

---

[38]ADA, Title I, Section 101(8).

[39]EEOC Interpretive Rules, supra, note 9.

[40]Id.

[41]ADA, Section 103(b).

[42]EEOC Interpretive Guidelines.

[43]Id.

[If] an individual poses a direct threat as a result of a disability, the employer must determine whether a reasonable accommodation would either eliminate the risk or reduce it to an acceptable level. If no accommodation exists that would either eliminate the risk or reduce the risk, the employer may refuse to hire an applicant or may discharge an employee who poses a direct threat.[44]

Title I of the ADA additionally provides that if a qualified fire service organization does not make reasonable accommodation for the known limitations of a qualified individual with a disability, it is considered to be discrimination. Only if the qualified fire service organization can prove that providing the accommodation would place an undue hardship on the operation of the fire service's business can discrimination be disproved.

Section 101 (9) defines a "reasonable accommodation" as

(a) making existing facilities used by employees readily accessible to and usable by the qualified individual with a disability" and includes (b) job restriction, part-time or modified work schedules, reassignment to a vacant position, acquisition or modification of equipment or devices, appropriate adjustments or modification of examinations, training materials, or policies, the provisions of qualified readers or interpreters and other similar accommodations for . . . the QID.[45]

The EEOC further defines "reasonable accommodation" as any one of the following:

1. Any modification or adjustment to a job application process that enables a qualified individual with a disability to be considered for the position such qualified individual with a disability desires, and which will not impose an undue hardship on the . . . business

---

[44]56 Fed. Reg. 35,745 (July 26, 1991); Also see, *Davis v. Meese*, 692 F. Supp. 505 (ED Pa. 1988)(Rehabilitation Act decision).
[45]ADA Section 101 (9).

2. Any modification or adjustment to the work environment, or to the manner or circumstances that the position held or desired is customarily performed, that enables the qualified individual with a disability to perform the essential functions of that position and which will not impose an undue hardship on the . . . business

3. Any modification or adjustment that enables the qualified individual with a disability to enjoy the same benefits and privileges of employment that other employees enjoy and does not impose an undue hardship on the . . . business[46]

In essence, the fire service organization is required to make "reasonable accommodations" for any/all known physical or mental limitations of the qualified individual with a disability unless the fire service organization can demonstrate that the accommodations would impose an "undue hardship" on the business or the particular disability directly affects the safety and health of the qualified individual with a disability or others. Included under this section is the prohibition against the use of qualification standards, employment tests, and other selection criteria that tend to screen out individuals with disabilities, unless the fire service organization can demonstrate the procedure is directly related to the job function. In addition to the modifications to facilities, work schedules, equipment and training programs, fire service organizations must initiate an "informal interactive (communication) process" with the qualified individual to promote voluntary disclosure of specific limitations and restrictions by the qualified individual to enable the fire service to make appropriate accommodations to compensate for the limitation.[47]

Job restructuring within the meaning of Section 101(9)(B) means modifying a job such that a disabled individual can perform its essential functions. This does not mean, however, that the essential

---

[46]EEOC Interpretive Guidelines.
[47]Id.

functions themselves must be modified.[48] Examples of job restricting may include:

- eliminating nonessential elements of the job
- redelegating assignments
- exchanging assignments with another employee
- redesigning procedures for task accomplishment
- modifying the means of communication that are used on the job[49]

Section 101 (10)(a) defines "undue hardship" as "an action requiring significant difficulty or expense," when considered in light of the following factors:

- nature and cost of the accommodation
- the overall financial resources and workforce of the facility involved
- the overall financial resources, number of employees, and structure of the parent entity
- the type of operation including the composition and function of the workforce, the administration and fiscal relationship between the entity and the parent[50]

The EEOC has proposed rules providing the factors of overall size of the operation, the structure and function of the workforce, the geographical location, and the "impact of the accommodation upon the operation of the site, including the impact on the ability of the other employees to perform their duties and the impact on the site's ability to conduct business" be considered.[51]

[48]See, *Gruegging v. Burke*, 48 Fair Empl. Prac. Cas. (BNA) 140 (DDC 1987); *Bento v. ITO Corp.*, 599 F. Supp. 731 (DRI 1984).

[49]EEOC Interpretive Guidelines, 56 Fed. Reg. 35,744 (July 26, 1991); Also see Rehabilitation Act decisions including *Harrison v. March*, 46 Fair Empl. Prac. Cas. (BNA) 971 (WD Mo. 1988); *Wallace v. Veteran Admin.*, 683 F. Supp. 758 (D. Kan. 1988).

[50]ADA Section 101(10)(a).

Section 102 (c)(1) of the ADA prohibits discrimination through medical screening, employment inquiries, and similar scrutiny. Underlying this section was Congress's conclusion that information obtained from employment applications and interviews "was often used to exclude individuals with disabilities, particularly those with so-called hidden disabilities such as epilepsy, diabetes, emotional illness, heart disease and cancer, before their ability to perform the job was even evaluated."[52] This section expanded regulations issued under the Rehabilitation Act related to pre-employment inquiries.[53]

Under Section 102(c)(2), fire service organizations are generally *Prohibited* from conducting pre-employment physical examinations of applicants and are also prohibited from asking the prospective employee if he/she is a qualified individual with a disability. Qualified fire service organizations are further *Prohibited* from inquiring as to the nature or severity of the disability even if the disability is visible or obvious. But fire service organizations may ask any candidates for transfer or promotion who have a known disability whether he/she can perform the required tasks of the new position if the tasks are job related and consistent with business necessity. The fire service organization is also permitted to inquire as to the applicant's ability to perform the essential job functions prior to employment. The fire service organization should use the written job descriptions as evidence of the essential functions of the position.[54]

Fire service organizations may require medical examinations *only* if the medical examination is specifically job related and is consistent with business necessity. Medical examinations are permitted *only after* the applicant with a disability has been *offered* the job position. The medical examination may be given before the

---

[51]EEOC Interpretive Guidelines.

[52]S. Comm. on Lab. and Hum. Resources Rep. at 38; H. Comm. on Jud. Rep. at 42.

[53]OFCCP Reg., 41 CFR Section 60-741.6(c). Seealso, *OFCCP v. EE Black, Ltd.*, 19 Fair Empl. Prac. Cas. (BNA) 1642 (DOL 1979), aff'd, 23 Fair Empl. Prac. Cas. (BNA) 1254 (D. Haw. 1980).

[54]ADA, Title I, Section 102(C)(2).

applicant *starts* the particular job and the job offer *may be condi-tioned* on the results of the medical examination *if* all employees are subject to the medical examinations and information obtained from the medical examination is maintained in separate confiden-tial medical files. Fire service organizations are permitted to con-duct voluntary medical examinations for current employees as part of an on-going medical health program but again the medical files must be maintained separately and in a confidential manner.[55]

The ADA does *not* prohibit fire service organizations from mak-ing inquiries or requiring medical examinations or "fit for duty" examinations when there is a need to determine whether an employee is still able to perform the essential functions of the job or where periodic physical examinations are required by medical standards or federal, state, or local law.[56]

The fire service organizations may test job applicants for alcohol and controlled substances *prior* to an offer of employment under Section 104 (d). This testing procedure for alcohol and illegal drug use is *not* considered a medical examination as defined under the ADA. Fire service organizations may additionally prohibit the use of alcohol and illegal drugs in the work place and may require that employees not be under the influence while on the job. Fire service organizations are permitted to test for alcohol and controlled sub-stance use by current employees in their workplace to the limits permitted by current federal and state law. The ADA requires all employers to conform to the requirements of the Drug-Free Work-place Act of 1988. Thus, most existing pre-employment and post employment alcohol and controlled substance programs that are not part and parcel of the pre-employment medical examination or on-going medical screening program will be permitted in their cur-rent form.[57]

---

[55]ADA Section 102(c)(2)(A).

[56]EEOC Interpretive Guidelines, 56 Fed. Reg. 35,751 (July 26, 1991). Federally mandated periodic examinations include such laws as the Rehabilitation Act, Occupational Safety and Health Act, Federal Coal Mine Health Act, and numerous transportation laws.

[57]ADA Section 102(c).

Individual employees who choose to use alcohol and illegal drugs are afforded *no protection* under the ADA, however, employees who have successfully completed a supervised rehabilitation program and are no longer using or addicted are offered the protection of a qualified individual with a disability under the ADA.[58]

Title II of the ADA is designed to prohibit discrimination against disabled individuals by public entities. This Title covers the provision of services, programs, activities, and employment by public entities. A public entity under Title II includes:

- a state or local government
- any department, agency, special purpose district or other instrumentality of a state or local government
- the National Railroad Passenger Corporation (Amtrak), and any commuter authority as this term is defined in section 103(8) of the Rail Passenger Service Act[59]

Title II of the ADA prohibits discrimination in the area of ground transportation including buses, taxis, trains and limousines. Air transportation is excluded from the ADA but is covered under the Air Carriers Access Act.[60] Covered organizations may be affected in the purchasing or leasing of new vehicles and in other areas such as the transfer of disabled individuals to the hospital or other facilities. Title II requires covered public entities to ensure that new vehicles are accessible to and usable by the qualified individual including individuals in wheelchairs. Thus, vehicles must be equipped with lifts, ramps, wheelchair space and other modifications unless the covered public entity can justify that such equipment necessary is unavailable despite a good faith effort to purchase or acquire this equipment. Covered organizations may want to consider alternative methods to accommodate the qualified individual such as use of an ambulance services or other alternatives.

---

[58]ADA Section 511(b).

[59]ADA Section 201(1).

[60]S. Rep 101-116, 21; H. Rep 101-485, Part 2; Part 3, 26-27.

Title III of the ADA builds upon the foundation established by the Architectural Barriers Act and the Rehabilitation Act. This Title basically extends the prohibitions that currently exist against discrimination in facility construction or financed by the federal government to apply to all privately operated public accommodations. Title III is focused on the accommodations in public facilities including such covered entities as retail stores, law offices, medical facilities and other public areas. This section requires that goods, services, and facilities of any public place be provided "in the most integrated setting appropriate to the needs of the (qualified individual with a disability)" except where the qualified individual with a disability may pose a direct threat to the safety and health of others that cannot be eliminated through modification of company procedures, practices, or policies. Prohibited discrimination under this section includes prejudice or bias against the qualified individual with a disability in the "full and equal enjoyment" of these services and facilities.[61]

The ADA makes it unlawful for public accommodations not to remove architectural and communication barriers from existing facilities and transportation barriers from vehicles "where such removal is readily achievable.[62] This statutory language is new and is defined as "easily accomplished and able to be carried out without much difficulty or expense."[63] As an example, moving shelves to widen an aisle, lowering shelves to permit access. The ADA also requires that when a commercial facility or other public accommodation is undergoing a modification that affects the access to a primary function area, specific alterations must be made to afford accessibility to the qualified individual with a disability.

Title III also requires "auxiliary aids and services" be provided for the qualified individual with a disability including, but not lim-

---

[61] ADA Section 302.
[62] ADA Section 302(b)(2)(A)(iv).
[63] ADA Section 301 (9).

ited to, interpreters, readers, amplifiers, and other devices (*not* limited or specified under the ADA) to provide the qualified individual with a disability with an equal opportunity for employment or promotion.[64] Congress did, however, provide that auxiliary aids and services need not be offered to customers, clients and other members of the public if the auxiliary aid or service creates an undue hardship on the business. Business and professional firms may use alternative methods of accommodating the qualified individual with a disability. This section also addresses modification of existing facilities to provide access to the qualified individual with a disability and requires all new facilities to be readily accessible and usable by the qualified individual with a disability.

Title IV requires all telephone companies to provide "telecommunications relay service" to aid the hearing and speech impaired qualified individual with a disability. The Federal Communication Commission has issued a regulation requiring implementation of this requirement by July 26, 1992 and has established guidelines for compliance. This section also requires that all public service programming and announcements funded with Federal monies be equipped with closed caption for the hearing impaired.[65]

Title V assures that the ADA does not limit or invalidate other federal or state laws that provide equal or greater protection for the rights of individuals with disabilities. Unique features of Title V include the miscellaneous provisions and the requirement of compliance to the ADA by all members of Congress and all federal agencies. Additionally, Congress required all state and local governments to comply with the ADA and permitted the same remedies against the state and local governments as any other organization.[66]

---

[64]ADA Section 3(1).

[65]Report of the House Committee on Energy and Commerce on the Americans With Disabilities Act of 1990, HR Rep. No. 485, 101st Cong., 2d Sess., (1990)(hereinafter cited as H. Comm. on Energy and Comm. Rep.); H. Comm. on Educ. and Lab. Rep., supra.; S. Comm. on Lab. and Hum. Resources Rep., supra.

[66]ADA Section 501.

Congress expressed their concern that sexual preferences could be perceived as a protected characteristic under the ADA or the courts could expand ADA's coverage beyond Congress's intent. Accordingly, Congress included Section 511 (b) which contains an expansive list of conditions that are *not* to be considered within the ADA's definition of disability. This list includes transvestites, homosexuals, and bisexuals. Additionally, the conditions of trans-sexualism, pedophilia, exhibitionism, voyeurism, gender identity disorders *not resulting from physical impairment* and other sexual behavior disorders are not considered as a qualified disability under the ADA. Compulsive gambling, kleptomania, pyromania, and psychoactive substance use disorders from *current illegal drug use* are also not afforded protection under the ADA.[67]

Individuals extended protection under this section of the ADA include all individuals associated with or having a relationship to the qualified individual with a disability. This *inclusion* is unlimited in nature, including family members, individuals living together, and an unspecified number of others.[68] The ADA extends coverage to all "individuals" thus the protection is provided to all individuals, legal or illegal, documented or undocumented, living within the boundaries of the United States regardless of their status.[69] Under Section 102 (b)(4), unlawful discrimination includes "excluding or otherwise denying equal jobs or benefits to a qualified individual because of the known disability of the individual with whom the qualified individual is known to have a relationship or associa-tion."[70] Thus, the protections afforded under this section are not limited to only family relationships, there appear to be no limits on the kinds of relationships or associations afforded protection. Of

---

[67]ADA, Section 511(a),(b); Section 508. There is some indication that many of the conditions excluded from the disability classification under the ADA may be considered a covered handicap under the Rehabilitation Act. See, *Rezza v. US Dept. of Justice*, 46 Fair Empl. Prac. Cas. (BNA) 1336 (ED Pa. 1988)(compulsive gambling); *Fields v. Lyng*, 48 Fair Empl. Prac. Cas. (BNA) 1037 (D. Md. 1988)(kleptomania).

[68]ADA Section 102(b)(4) and 302(b)(1)(E).

[69]H. Rep. 101-485,Part 2, 51.

[70]ADA Section 102(b)(4).

particular note is the *inclusion* of unmarried partners of persons with AIDS or other qualified disabilities under this section.[71]

As with most regulatory legislation, the ADA requires that employers *post* notices of the pertinent provisions of the ADA in an accessible format in a conspicuous location within the employer's facilities. A prudent organization may wish to provide additional notification on their job applications and other pertinent documents.[72]

Under the ADA, it is unlawful for an employer to "discriminate on the basis of disability against a qualified individual with a disability" in all areas including:

1. recruitment, advertising, and job application procedures
2. hiring, upgrading, promotion, award of tenure, demotion, transfer, layoff, termination, right to return from layoff, and rehiring
3. rate of pay or other forms of compensation and changes in compensation
4. job assignments, job classifications, organization structures, position descriptions, lines of progression, and seniority lists
5. leaves of absence, sick leave, or other leaves
6. fringe benefits available by virtue of employment, whether or not administered by the employer
7. selection and financial support for training including apprenticeships, professional meetings, conferences and other related activities, and selection for leave of absence to pursue training
8. activities sponsored by the employer including social and recreational programs
9. any other term, condition, or privilege of employment[73]

The EEOC has also noted that it is "unlawful ... to participate in a contractual or other arrangement or relationship that has the

---

[71]H. Rep. 101-485, Part 2, 61-62; Part 3, 38-39.

[72]ADA Section 105.

[73]EEOC Interpretive Guidelines.

effect of subjecting the covered entity's own qualified applicant or employee with a disability to discrimination." This prohibition includes referral agencies, labor unions (including collective bargaining agreements), insurance companies and others providing fringe benefits, and organizations providing training and apprenticeships.[74]

The ADA requires equal access for the qualified individual with a disability to whatever health insurance coverage the fire service organization provides to all employees. However, the ADA does *not* affect pre-existing condition clauses in insurance policies, so long as the clause does not subterfuge to evade the purposes behind the ADA.[75] This provision may be applicable to fire service organizations in the area of health insurance, disability, and other insurance contracts.

Other notable provisions of the ADA, or lack thereof, include no recordkeeping requirements, no affirmative action requirements, and no preclusions or restrictions on smoking in the place of employment. The ADA has no retroactivity provisions.

Congress gave the ADA with the same enforcement and remedies as Title VII of the Civil Rights Act of 1964 and have included the remedies provided under the Civil Rights Act of 1991. As with Title VII, compensatory and punitive damages (with upper limits) have been added as remedies in cases of intentional discrimination, and there is a correlative right to a jury trial. Unlike Title VII, there is an exception where there exists a good faith effort at reasonable accommodation.[76]

For now, the enforcement procedures adopted by the ADA mirror those of Title VII of the Civil Rights Act. A claimant under the ADA must file a claim with the EEOC within 180 days from the alleged discriminatory event or within 300 days in states with approved enforcement agencies such as the Human Rights Com-

---

[74]Id.

[75]ADA Section 501(c). See Also, Senate Report accompanying S933.

[76]Civil Rights Act of 1991, Section 102.

mission. The EEOC has 180 days to investigate the allegation and to sue the employer or issue a right-to-sue notice to the employee. The employee will have 90 days to file a civil action from the date of this notice.[77]

The original remedies provided under the ADA included reinstatement, with or without back pay, and reasonable attorney fees and costs. The ADA also provided for protection against retaliation against the employee for filing the complaint or others who may assist the employee in the investigation of the complaint. The ADA remedies are designed, as with the Civil Rights Act, to make the employee "whole" and to prevent future discrimination by the employer. All rights, remedies, and procedures of Section 505 of the Rehabilitation Act of 1973 are also incorporated into the ADA. Enforcement of the ADA is also permitted by the Attorney General or by private lawsuit. Remedies under these Titles included ordered modification of a facility, and civil penalties up to $50,000.00 for the first violation and $100,000.00 for any subsequent violations. Section 505 permits reasonable attorney fees and litigation costs for the prevailing party in an ADA action but, under Section 513, Congress encourages the use of arbitration to resolve disputes arising under the ADA.[78]

With the passage of the Civil Rights Act of 1991, the remedies provided under the ADA were modified. Damages for employment discrimination, whether intentional or by practice that has a discriminatory effect, may include hiring, reinstatement, promotion, back pay, front pay, reasonable accommodation, or other action that will make an individual "whole." Payment of attorneys' fees, expert witness fees and court courts were still permitted and jury trials were allowed.

Compensatory and punitive damages were also made available where intentional discrimination is found. Damages may be available to compensate for actual monetary losses, for future monetary

---

[77]S. Rep. 101-116, 21; H. Rep. 101-485 Part 2, 51; Part 3, 28.
[78]ADA Section 505 and 513.

losses, for mental anguish and inconvenience. Punitive damages area also available if an employer acted with malice or reckless indifference. The total amount of punitive damages and compensatory damages for future monetary loss and emotional injury for each individual is limited, based upon the size of the employer.

Number of Employees	Damages Will not Exceed
15–100	$50,000
101–200	$100,000
201–500	$200,000
500 or more	$300,000

Punitive damages are *not* available against state or local governments and thus related fire service organizations.[79]

FAIR LABOR STANDARDS ACT

The Fair Labor Standards Act (referred to as the FLSA) was enacted by congress in 1938.[80] The FLSA was amended in 1947 by the Portal-to-Portal Act,[81] in 1961 (adding "enterprise" definition), and in 1977 (tightened statutory exemptions). The Equal Pay Act of 1963 also amended the FLSA by forbidding employers and unions from differentiating wages and benefits based on sex.[82] The 1974 amendment to the FLSA extended coverage to employees of federal, state, county, and municipal governments.[83]

The FLSA, as amended, generally required the payment of specified minimum wages and overtime pay for hours worked over forty in any one week prior of time. Preliminary and postliminary activities, such as coming and going to work, are generally excluded under the Portal-to-Portal exclusion. "Working time" is

---

[79]The EEOC has published additional guidance documents regarding the ADA. Contact your local EEOC office to acquire copies of these Regulations and Interpretations.

[80]29 U.S.C.A. §201 et. seq.

[81]29 U.S.C.A. § 251 et. seq.

[82]29 U.S.C.A. § 206.

[83]29 U.S.C.A. § 203(5)(6). Also see, *Garcia v. San Antonio Met. Transit Auth.*, 26 W&H Cases 65 (S.Ct. 1985).

defined as the time for which an employee is entitled to compensation and includes all time an employee is required to be on duty on the employer's premises or at a prescribed work place.[84] In the fire service, most fire fighters will be considered employees and thus covered under the FLSA. Exceptions from the requirements exists in the following areas:

1. executive exemption
2. administrative exemption
3. professional exemption
4. outside salesperson exemption

Fire service organizations should be aware that specialized tests have been designed to evaluate whether an individual or position "fits" the exemption category. In general, fire service officers usually fall within one or more exemption categories.

As a general rule, all hours that a fire fighter is required to provide the fire service organization are considered compensable work time. Additionally, rest and meal periods, on-duty wait time, on-call wait time, reporting time, sleep time, training time, civic and charity work (if done at the employer's request), and travel time can also be compensable. Clothes changing and wash-up activities can be excluded[85] as well as pre-employment tests, personal medical attention, voluntary training programs, athletic events, voluntary civic or volunteer work, illness time, leave of absence, vacations, and holidays.

The Wage and Hour Administration has permitted the reasonable cost or fair value of certain noncash items to be included in the computation of wages for the purposes of satisfying the minimum wage requirements under the FLSA. These items may include meals furnished for the benefit of the fire fighter, lodging, merchandise, transportation, tuition, savings bonds, insurance premiums, and union dues.

---

[84]29 CFR § 778.223.

[85]*Mitchell v. Southeastern Carbon Paper Co.*, 228 F.2d 934 (CA_1955).

GARNISHMENTS

In recent years, a greater importance has been placed upon personal privacies in the area of credit reporting and other laws in the area of collections. The Federal Wage Garnishment Law sets restrictions on the amount of employee earnings that can be deducted within a given period and offers protection from discharge from employment for reasons of garnishment. Garnishment is the process by which a person's debts are collected in order to fulfill a debt to a third party. The most frequent garnishments in the average workplace are for child support or alimony payments. As an example, if an individual has been awarded child support by the court for X dollar amount and the party who was supposed to pay the child support has neglected to do so for a period of time, a garnishment action may be brought where the individual owing child support may have a percentage of his/her wages taken and held by the employer and forwarded to the courts or the third party on a periodic basis.

Under the law, when a garnishment order has been issued for the support of any person, no more than 50 percent of disposable income of an individual supporting a former spouse or dependant child can be garnished, and no more than 60 percent of the individual's wages are to go to support of a second wife and dependant child. An additional 5 percent may be held in each situation if there are outstanding arrear agents more than 12 weeks old. Wage garnishment law also sets forth the amount of money that can be garnished from an individual's wages. Where an employee's disposable earnings (i.e., wages remaining after deductions required by law) equals more than $134 a week, up to 25 percent of these earnings may be garnished to pay the prior debt. When disposable earnings are less than $134, the amounts over $100.50 may be garnished. It should be noted that several states have placed further limits on the amounts and the types of earnings that can be garnished. Minimum wage garnishment law is only considered to be a minimum level of protection of individuals and additional state laws may be applied to add further protection against garnishment or, on the other hand, provide greater access to earnings depending upon the situation.

SICK PAY/OVERTIME

In the public sector, absence from work because of illness usually does not deprive an individual of his/her salary or compensation. Specifically, public sector employers normally continue to pay the officer and hold the position open through any reasonable period of time in which the individual is ill. During a period of illness, most public sector fire service organization employees are normally provided with sick pay as set forth by either state statute, a collective bargaining agreement or organization policies. It should be noted that a fire service organization is not required by law to guarantee sick pay. There is no requirement under the Fair Labor Standards Act or any other federal legislation that requires a fire service organization to pay sick pay.

Sick pay may vary greatly depending upon the fire service organization, the type of health benefits provided, and even to the type of injury involved. In many fire service organizations, if an individual is injured while on the job, the fire service organization will continue to pay the fire fighters salary instead of the fire fighter collecting a lesser amount under worker's compensation. The fire service organization would be reimbursed for the 66 2/3 percent of the worker's salary under worker's compensation and would continue to contribute the additional amounts in order that the fire fighter would continue to collect his normal salary. For injuries incurred outside of the job, fire service organizations often provide health care benefits that pay the fire fighter's salary during the period of illness.

For long-term or catastrophic illnesses or injuries, fire service organizations have a designated time in which the individual is maintained on salary with benefits. When an individual has incurred an injury that may not permit him/her to return to the fire service organization in their entire capacity, many fire service organizations have looked at alternatives such as restrictive duty programs, re-training for the fire fighter, and even early retirement programs. In the area of overtime, the landmark case of *Garcia v. San Antonio Metropolitan Transit Authority,* 105 S.Ct. 1005, 83 L.ed. 2d 1016 (1985), the United States Supreme Court held that local employment relations are subject to the standards of the Fair Labor Standards Act (FLSA). As discussed above, the FLSA sets forth a basic minimum wage and overtime pay standards for public employees, including fire fighters and police officers. In general, the FLSA directs that each employee be paid for "all hours worked." This includes all time in which the fire fighter might be on duty, on the employer's premises or at the prescribed workplace of the employer, or performing duties directed by the fire service organization. In fire service organizations, the unique shift arrangement, which combines active duty, rest, standby and sleep periods, requires a special computation arrangement provided under the FLSA for shift personnel.

Section 7 (K) of the Fair Labor Standards Act provides partial exemptions of the requirements to pay a fire fighter overtime of hours worked beyond 40 hours per calendar week. Under this provision, a public employee (i.e., state or local government agency) can establish a work period of fewer than 28 consecutive days, but not less than 7 consecutive days, in lieu of the standard 40 hour work week. The work period is the basis of computing any required payment for overtime. The rule provides that where fire fighters work time does not exceed 212 hours in any period of 28 consecutive days, there is no requirement for payment of overtime. Work periods can stretch from 7 to 28 days, with each period assigned a maximum number of work hours. Fire service organizations should be very knowledgeable in the specific scheduling rules set forth under Section 7 (K). Many fire service organizations should note that any deviation from these specific scheduling rules may subject them to retrospective enforcement (i.e., back pay) but note that there is a two year statute of limitations on recovery. Others should note that they can not waive their rights under the FLSA, and no contract or collective bargaining agreement can waive the overtime provisions as required by law. Fire service organizations cannot simply announce that overtime will not be paid at all or unless authorized in advance. The employee retains the right to seek the overtime wages due at any time that he/she believes that improper payment of wages have been made. FLSA overtime provisions are only the minimum requirements. Several state statutes and local agreements can provide greater overtime benefits than required under FLSA. The FLSA is the "bare bones" minimum requirement in which a fire service organization must comply in the area of payment of wages. Let it be noted, that the Fair Labor Standards Act provides a special exemption from the rules for fire service organizations with fewer than five full-time and part-time employees. However, despite the fact of this exemption, the fire service organization may still be subject to overtime provisions under state statutes or local ordinances or agreements.

FAMILY AND MEDICAL LEAVE ACT

One of the newest laws to impact may fire service organizations is the Family and Medical Leave Act of 1993 (see Table 9.1).[86] The key provisions of this law for fire service organizations are Title I and Title II. In Title I, the definition of eligible employee (i.e., individuals provided coverage under this law) is defined as being an employee who has been employed at least 12 months with the fire service organization and who has provided at lease 1,250 hours of service during the 12 months. Fire service organizations with fewer than 50 employees within a 75 mile radius of the worksite are excluded from coverage.

Title I also includes the broadly construed definitions of the terms "parent," "son or daughter," (includes biological, adopted or

---

[86]29 U.S.C. § 2601-2654.

TABLE 9.1   State Family Leave Laws/Rules and Provisions

	Employees Covered	Maximum Leave		Employees Covered	Maximum Leave
Alabama	○		Nebraska		
Alaska	○		Nevada		
Arizona			New Hampshire		
Arkansas			New Jersey	▲	12 wk per 24 mo
California	▲	16 wk per 24 mo	New Mexico		
Colorado			New York		
Connecticut	▲	16 wk per 24 mo	North Carolina		
Delaware	○[a]		North Dakota	○	
D.C.	▲	16 wk per 24 mo	Ohio		
Florida	○		Oklahoma	○	
Georgia	○		Oregon	▲	12 wk per 24 mo
Hawaii	▲[c]	4 wk per 12 mo	Pennsylvania		
Idaho	○		Puerto Rico		
Illinois	○		Rhode Island	▲	13 wk per 24 mo
Indiana			South Carolina	○	
Iowa	○		South Dakota		
Kansas			Tennessee		
Kentucky	▲[a]		Texas		
Louisiana	○		Utah		
Maine	▲	10 wk per 24 mo	Vermont	▲	12 wk per 12 mo
Maryland	○		Virginia		
Massachusetts			Virgin Islands		
Michigan			Washington	▲	12 wk per 24 mo
Minnesota	▲	6 wk per 12 mo	West Virginia	○	
Mississippi			Wisconsin	▲	2 wk per 12 mo[b] 6 wk per 12 mo[d]
Missouri	○[a]		Wyoming		
Montana					

○ Public employees only   ▲ Public and private employees
[a]Provides adoption leave only
[b]Provides leave for medical reasons only
[c]Effective as of 1994
[d]Provides leave for adoption or birth of a child only

foster child or legal ward), but does not define "spouse." "Serious health condition" is defined as "an illness, injury, impairment, or physical or mental condition" involving either inpatient care or continuing treatment by a health care provider.

Eligible fire fighters are entitled to 12 unpaid work-weeks of leave during any 12-month period for three basic reasons: (1) the birth or placement for adoption or foster care of child (within 12 months of the birth or placement); (2) serious health condition of a spouse, child, or parent; or (3) the fire fighter's own serious health condition.

In general, the fire service organization may require that the eligible fire fighter provide certification of a serious health condition of himself/herself or a family member. Certification includes the date of the serious health condition, the duration of the condition, appropriate facts regarding the condition, and a statement that the eligible fire fighter needs to care for the spouse, child, or other family member. For cases of intermittent leave, the dates of the leave should also be noted. The fire service organization may require a second opinion be obtained by the fire fighter at his/her own expense. The second opinion may not be provided by the health care provider employed by the fire service organization. In the event of a conflict in opinions, a third and final opinion can be required at the expense of the fire service organization.

The fire fighter who completes a period of leave is to be returned either to the same position or to a position equivalent in pay, benefits, and other terms and conditions of employment. This leave cannot result in the loss of any previously accrued seniority or employment benefits but neither of these benefits are required to accrue during the leave. Health benefits are to continue throughout the leave. The fire service organization may be required to pay the health coverage premiums but may recover these premiums if the fire fighter fails to return to the job following the leave. As with most laws, the fire service organization is prohibited from discriminating against the fire fighter for use of the Family and Medical Leave Act. The agency vested with the authority to enforce the FMLA is the U.S. Department of Labor. The powers to investigate, prosecute, maintain records, and enforce are similar to the Fair Labor Standards Act. A Commission on Leave has also been established on a temporary basis to study the impact of the FMLA. The statute of limitations on filing an action under the FMLA is 2 years from the last event and 3 years if the violation is willful. For fire fighters in the federal civil service system, Chapter 63 of Title 5, United States Code, extends coverage to federal civil service employees.

Fire service organizations should be aware that the FMLA is the foundation for coverage in this area and states or municipalities may provide benefits equal to or greater than the FMLA. As can be seen from the analysis below, 35 states have enacted varying forms of the FMLA. Nothing in the FMLA is meant to discourage fire service organizations from offering more generous leave policies.

WORKERS COMPENSATION

In virtually every state, all paid fire fighters are provided protection from related injuries by statute. In some states, volunteer fire fighters are additionally afforded protection by statute for work related injuries. It should be noted that each state has its own specific workers compensation laws and the variation between the states may be significant. The basic concept behind workers compensation is to compensate monetarily the injured fire fighter who sustained an injury or illness that arose from the course of his/her employment.

In most states, there are three basic components of the workers compensation system. The first basic component consists of payment of medical bills. As a rule, all medical expenses incurred as the result of a work related injury or illness are compensable under workers compensation. This category includes doctor's fees, hospital fees, rehabilitation fees, and other expenses incidental to the treatment of the injury or illness. In some states, this coverage may be extended to such items as modification of vehicles, modification of home, and replacement of medical devices. A component of most workers compensation systems consists of payment for wages while the employee is not able to work. This is often called time loss benefits or permanent partial disability benefits (PPD). The general rule, the injured employee would receive 66 2/3 percent of his salary with a minimum and a maximum provided by statute during the period for which they are off work. For example, in state X, the employee would receive 66 2/3 percent of the average weekly wage of when he/she was off work. If an employee were making $500.00 per week, he could receive benefits of $333 per week while recuperating from the injury. If an employee were making $50.00 per week, the 66 2/3 percent would provide approximately $33 per week in time loss benefits. However, in most states, a maximum amount to be received has been established by statute in addition to a minimum amount that can be received for time loss benefits. If 66 2/3 percent of the injured employee's salary was over the maximum amount, the employee would only receive the maximum permissible amount. Conversely, if the employee were under the minimum amount, the employee would receive the minimum amount by statute. Time loss benefits or permanent partial disability benefits are normally tax free to the injured employee.

The third component of most workers compensation systems includes compensation to the employee for the permanently incurred injury. In some states this is called a permanent total rating. In this rating system, each part of the body has been provided an amount of weeks or monies total disability because of the work related injury or illness. When the employee has reached maximum medical recovery, the attending physician or panel of physicians will rate the amount of disability the individual has incurred because of the injury or illness. This amount of disability is then calculated according to the body part and type of injury and the amount of disability the individual has incurred. This percentage of disability is then calculated in accordance with the schedule established in the state. The employee is then provided a set amount, whether in lump sum or in payments, for that injury. For example, if a fire fighter should have his/her hand amputated at the wrist, this would constitute a 100 percent disability of the hand. When the fire fighter reaches maximum medical recovery, he/she would receive a set amount of money as determined by the state statute or by the state workers compensation laws for whatever the state has determined that hand would be worth. Let it be noted that the amounts established by state statute vary from state to state.

In most states, the fire service organization is required by law to acquire workers compensation coverage. In some states, the fire service organization or any employer will be provided a number of different options in which to acquire this coverage. Some states offer workers compensation through a state agency while other states permit the employer or fire service organization to acquire workers compensation through an established insurance company. Most states permit larger employers to become "self-insured." By self-insured, the employer would basically pay for all medical costs, time loss benefits, and permanent partial disability rating amounts from their bottom line. To apply for self-insurance coverage, the employer normally has to post a bond with the state, to let them know that they are able to properly administer the workers compensation system. Many employers hire outside contractors to administer their workers compensation program for them.

The compensation is normally the sole remedy for any work related injuries. To qualify for a work related injury or illness, the injury or illness must have arisen out of or in the course of the employment. Many states have established an administrative procedure through which particular forms are completed for the employee to initiate workers compensation benefits. If workers compensation benefits have been provided, the injured employee normally waives any rights to common law suits against the employer for the injury or illness. Let it also be noted that employees may not, under normal circumstances, waive their right to workers compensation. In some states, prior to an injury or illness, an employee may waive his right to workers compensation by providing a written waiver to the state's workers compensation commission. Although attorneys are not required for workers compensation coverage, most states have established a set fee for attorneys representing employees for workers compensation claims.

In essence, workers compensation is a method by which an employee and his/her family are relieved of the financial burden following a work related injury or illness while protecting fire service organization from the potential million-dollar judgements. The workers compensation laws normally are to be liberally construed in favor of the employee. Administrative procedures have been established under most workers compensation systems that permit the employee easy access to the benefits provided by law and quick payment of these benefits. These responsibilities are provided by statute to both the injured fire fighter and the fire service organization. The fire service organization, in most states, has the responsibility of providing the benefits to the injured fire fighter but also has the right to evaluate medical records, investigate the particular incident, and, if found not to be work related, to deny the claim by the fire fighter.

SCREENING AND MEDICAL EXAMINATIONS

In the event of pre-employment screening, post-employment screening and other related testing of employees on the job has become a very controversial issue in recent years. As noted above in the American with Disabilities Act Section, a law may specify what type of testing is permitted and when in the hiring process an employee may be tested. Fire service organizations should be aware that there are not only federal laws addressing pre- and post-employment screening but also state laws and even local ordinances regulate these areas.

The hiring or promotion process in most fire service organizations is one of the most important processes and is often one of the least understood in the relationship between the fire fighter and fire service organization. Hiring and promotion decisions are often based on information acquired through screening devices and procedures and the decisions are made behind closed doors. Unsuccessful candidates are often not provided the results of the screening process nor are they provided the reasons why he/she was not hired or promoted. Thus, with the process cloaked in secrecy, the screening and hiring/promotion process can often be vulnerable to challenge by unsuccessful candidates.

The base level screening process used by virtually all fire service organizations is the employment application. Simply asking an inappropriate or unlawful question on the employment application form may subject the fire service organization to potential liability. Table 9.2 shows some of the considerations that should be evaluated on the employment application.[87]

Another area in the screening process with potential dangers for fire service organizations is the interview process. As with the application itself, the interviewer is prohibited from making specific inquiries that may be discriminatory, such as, "Tell me how you can be a fire fighter if you're in a wheelchair?" Careful preparation prior to a pre-employment or promotion interview should include a specified series of questions that have been evaluated to ensure that each question itself is not discriminatory on its face (See Section on Americans With Disabilities Act.)

The area of pre-employment medical screening is fast becoming a hotbed of potential legal risks for fire service organizations. As noted in the section on the Americans With Disabilities Act, pre-employment medical screening procedures, questionnaires, or evaluations are prohibited in most fire service organizations. Post employment medical screening and voluntary medical screening is permitted. Alcohol and drug testing is usually not considered part of the medical screening process and can be performed at any time in the pre-employment screening.

The types of medical screening are wide ranging from the basic "stick out your tongue and cough" to elaborate neuromuscular testing. Fire service organizations should closely evaluate the type and method of such medical screening to ensure the accuracy and non-discriminatory effects of this process.

---

[87]Rothstein, Knapp & Liebman, Employment Law, University Casebook (1987).

TABLE 9.2 Considerations for Employment Application

Subject	Lawful Pre-employment Inquiries	Unlawful Pre-employment Inquiries
Name	First, middle, last. Name used if previously employed under different name.	To require prefix to applicant's name (Mr., Mrs., Miss, Ms). Inquiry into marital status. Inquiry into previous name where it has been changed by court order.
Sex		Inquiry into sex of applicant.
Height and weight		It is unlawful for an employer to set minimum height or weight requirements for hiring unless based on a legitimate job need.
Address	Applicant's place of residence and length of residence.	Inquiry into foreign addresses which would indicate national origin.
Age	Are you under 18 or over 70? If there is a question as to applicant being of legal working age, proof may be requested in form of work permit.	Requesting an individual date of birth prior to employment is prohibited.
Character	Permissible to ask applicant for character references.	It is unlawful to inquire from references any information that is directly prohibited by West Virginia Human Rights Act.
Physical handicaps (also see ADA)	Have you any disability that would prevent you from performing the duties of the job for which you are applying? If yes, explain.	It is unlawful to discriminate against an individual who is blind if the individual is able and competent to perform the services required.
Economic status		It is inadvisable to inquire as to bankruptcy, car ownership, rental or ownership of a house, length of residence at an address, or past garnishment of wages as poor credit ratings have a disparate impact on women and minorities.
Number of dependents	This information may be requested for legitimate purposes only after hiring.	Asking an applicant's number of dependents prior to employment is prohibited.
Race or color		Any inquiry which would indicate race or color is prohibited.
Color of hair or eyes		Inquiry into color of hair or eyes is prohibited.
Photographs	Photographs may be requested only after hiring and then only for legitimate business purpose.	Any request for photographs prior to hiring is prohibited.

TABLE 9.2 (continued)

Subject	Lawful Hyperemployment Inquiries	Unlawful Pre-employment Inquiries
Religion, creed	May be asked only after hiring if employer informs applicant that it is to be used for emergency purposes only.	Inquiry into applicant's religious denomination, religious affiliations, church, parish, pastor, or religious holidays observed prior to hiring is prohibited.
Citizenship	Are you a citizen of the United States? If applicant is not a citizen, employer may require a work permit or evidence of alien status.	To ask if applicant is naturalized or a native-born citizen; or to ask the date applicant acquired citizenship. To require prior to hiring that applicant produce naturalization papers or first papers. To ask if applicant's spouse or parents are citizens of the United States.
Birthplace	Proof of citizenship may be requested after hiring.	Inquiry into birthplace of applicant, or birthplace of applicant's parents, spouse, or relatives. Require prior to hiring, birth certificate, naturalization or baptismal record.
National origin	To inquire what languages applicant reads, speaks, and writes fluently.	Inquiry into an applicant's lineage, ancestry, national origin, descent, parentage, or nationality. Nationality of parents or spouse. Inquiry into how applicant acquired ability to read, write, or speak a foreign language.
Education	Inquiry into what academic professional, or vocational schools attended.	It is unlawful to ask specifically the nationality, racial, or religious affiliation of a school attended by the applicant.
Prior arrest record		The requiring of arrest information has been shown to have a disparate effect on racial minorities.
Criminal record		Inquiry advisable only if job related.
Relatives	Inquiry into name and address and relationship of persons to be notified in case of emergency. This information may be solicited only after hiring.	Inquiry into the location of relatives' places of business. Inquiry to determine if relatives of applicant are or have previously been employed by the employer.
Military service	Inquiry into applicant's experience or duties in United States armed forces	To require copy of military discharge paper or military discharge number
Organizations	Inquiry into organization memberships, excluding those organizations which may indicate race, religion, color, sex, national origin, ancestry of their members	Unlawful to inquire into organizations which may indicate race, religion, color, sex, or ancestry of their members

Drug and alcohol screening is fast becoming a mainstay in the employment screening process for fire service organizations. Pre-employment testing is permitted in virtually all circumstances. Fire service organizations should take particular caution in the design of the testing program, the method of acquiring the sample, the chain of custody of the sample, and the actual testing of the sample. Inaccuracies in the testing procedure may be the basis for legal liability. Alcohol and drug testing for fire fighters currently employed can be used dependent on whether the fire service organization is permitted to use random testing or "for cause" testing.

With the enactment of the Employee Polygraph Protection Act,[88] the use of polygraph examinations as a screening mechanism was vastly curtailed in the fire service. This law basically prohibits private sector fire service organizations from using a polygraph examination at any time during the employment screening process and during employment with the fire service organization. Public fire service organizations who are part of state or local government, or any political subdivision, are exempt from this law.[89] Violations of this law can result in a private cause of action against the fire service organization.

The above are but a few of the laws that directly or indirectly affect the hiring process for fire service organizations. A prudent fire service organization should perform a careful analysis of their hiring and screening processes on a periodic basis to ensure compliance with all federal, state, and local law.

SELECTED CASES AND STATUTES

Family and Medical Leave Act

29 U.S.C. §§ 2601-2654

---

[88] 29 U.S.C. § 2001-2009.

[89] 29 U.S.C. § 2006. (Note: National defense, security, and FBI contractors are also exempt from this law).

SEC. 101—Definitions

As used in this Title:

1. Commerce—any activity, business, or industry in commerce or in which a labor dispute would hinder or obstruct commerce or the free flow of commerce, and include "commerce" and any "industry affecting commerce," as defined in paragraphs (1) and (3) of section 501 of the Labor Management Relations Act, 1947 (29 U.S.C. 142(1) and (3)).
2. Eligible employee
   a. in general—an employee who has been employed for at least 12 months by the employer with respect to whom leave is requested under section 102; and for at least 1,250 hours of service with such employer during the previous 12-month period.
   b. Exclusions—does not include any Federal officer or employee covered under Subchapter V of Chapter 63 of Title 5, United States Code (as added by Title II of this Act) or any employee of an employer who is employed at a worksite at which such employer employs less than 50 employees if the total number of employees employed by that employer within 75 miles of that worksite is less than 50.
   c. Determination—the legal standards established under section 7 of the Fair Labor Standards Act of 1938 (29 U.S.C. 207) shall apply.
3. Employ; employee, state—the same meanings given such terms in subsections (c), (e), and (g) of section 3 of the Fair Labor Standards Act of 1938 (29 U.S.C. 203(c), (e), and (g)).
4. Employer
   a. In general—any person engaged in commerce or in any industry or activity affecting commerce who employs 50 or more employees for each working day during each of 20 or more calendar workweeks in the current or preceding calen-

dar year and includes any person who acts, directly or indirectly, in the interest of an employer to any of the employees of such employer; any successor in interest of an employer; and includes any "public agency," as defined in section 3(x) of the Fair Labor Standards Act of 1938 (29 U.S.C. 203(x)).

   b. Public agency—a person engaged in commerce or in an industry or activity affecting commerce.

5. Employment benefits—all benefits provided or made available to employees by an employer, including group life insurance, health insurance, disability insurance, sick leave, annual leave, educational benefits, and pensions, regardless of whether such benefits are provided by a practice or written policy of an employer or through an "employee benefit plan," as defined in section 3(3) of the Employee Retirement Income Security Act of 1974 (29 U.S.C. 1002(3)).

6. Health care provider—either of:

   a. A doctor of medicine or osteopathy who is authorized to practice medicine or surgery (as appropriate) by the State in which the doctor practices

   b. Any other person determined by the Secretary to be capable of providing health care services.

7. Parent—the biological parent of an employee or an individual who stood in loco parentis to an employee when the employee was a son or daughter.

8. Person—the same meaning given such term in section 3(a) of the Fair Labor Standards Act of 1938 (29 U.S.C. 203(a)).

9. Reduced Leave Schedule—a leave schedule that reduces the usual number of hours per workweek, or hours per workday, of an employee.

10. Secretary—the Secretary of Labor

11. Serious Health Condition—an illness, injury, impairment, or physical or mental condition that involves

   a. inpatient care in a hospital, hospice, or residential medical care facility.

   b. continuing treatment by a health care provider.

12. Son or Daughter—a biological, adopted, or foster child, a step-child, a legal ward, or a child of a person standing in loco parentis, who meets one of the following criteria:
    a. Under 18 years of age.
    b. 18 years of age or older and incapable of self-care because of a mental or physical disability.
13. Spouse—a husband or wife, as the case may be.

SEC. 102—Leave Requirement

In General

1. Entitlement to leave—subject to Section 103, an eligible employee shall be enTitled to a total of 12 workweeks of leave during any 12-month period for one or more of the following:
    a. Because of the birth of a son or daughter of the employee and in order to care for such son or daughter.
    b. Because of the placement of a son or daughter with the employee for adoption or foster care.
    c. In order to care for the spouse, or a son, daughter, or parent, of the employee, if such spouse, son, daughter, or parent has a serious health condition.
    d. Because of a serious health condition that makes the employee unable to perform the functions of the position of such employee.
2. Expiration of entitlement—the entitlement to leave under sub-paragraphs (A) and (B) of paragraph (1) for a birth or place-ment of a son or daughter shall expire at the end of the 12-month period beginning on the date of such birth or placement.

Leave Taken Intermittently or on a Reduced Schedule
Leave under subparagraph (A) or (B) of subsection (aX1) shall not be taken by an employee intermittently or on a reduced leave schedule unless the employee and the employer of the employee agree otherwise. Subject to paragraph (2), subsection (e)(2), and

section 103(b)(5), leave under subparagraph (C) or (D) of subsection (a)(1) may be taken intermittently or on a reduced leave schedule when medically necessary. The taking of leave intermittently or on a reduced leave schedule pursuant to this paragraph shall not result in a reduction in the total amount of leave to which the employee is entitled under subsection (a) beyond the amount of leave actually taken.

3. Alternative position—if an employee requests intermittent leave, or leave on a reduced leave schedule, under subparagraph (C) or (D) of subsection (a)(1), that is foreseeable based on planned medical treatment, the employer may require such employee to transfer temporarily to an available alternative position offered by the employer for which the employee is qualified and
   a. Has equivalent pay and benefits
   b. Better accommodates recurring periods of leave than the regular employment position of the employee.

Unpaid Leave Permitted
Except as provided in subsection (d), leave granted under subsection (a) may consist of unpaid leave. Where an employee is otherwise exempt under regulations issued by the Secretary pursuant to section 13(a)(1) of the Fair Labor Standards Act of 1938 (29 U.S.C. 213(a)(1)), the compliance of an employer with this Title by providing unpaid leave shall not affect the exempt status of the employee under such section.

Relationship to Paid Leave
Unpaid Leave—If an employer provides paid leave for fewer than 12 workweeks, the additional weeks of leave necessary to attain the 12 workweeks of leave required under this Title may be provided without compensation.

4. Substitution of paid leave:

a. In general—an eligible employee may elect, or an employer may require the employee, to substitute any of the accrued paid vacation leave, personal leave, or family leave of the employee for leave provided under subparagraph (A), (B), or (C) of subsection (a)(1) for any part of the 12-week period of such leave under such subsection.

b. Serious health condition—an eligible employee may elect, or an employer may require the employee, to substitute any of the accrued paid vacation leave, personal leave, or medical or sick leave of the employee for leave provided under subparagraph (C) or (D) of subsection (a)(1) for any part of the 12-week period of such leave under such subsection, except that nothing in this Title shall require an employer to provide paid sick leave or paid medical leave in any situation in which such employer would not normally provide any such paid leave.

Foreseeable Leave

Requirement of Notice—In any case in which the necessity for leave under subparagraph (A) or (B) of subsection (a)(1) is foreseeable based on an expected birth or placement, the employee shall provide the employer with not less than 30 days' notice, before the date the leave is to begin, of the employee's intention to take leave under such subparagraph, except that if the date of the birth or placement requires leave to begin in less than 30 days, the employee shall provide such notice as is practicable.

5. Duties of employee—in any case in which the necessity for leave under subparagraph (C) or (D) of subsection (a)(1) is foreseeable based on planned medical treatment, the employee

a. Shall make a reasonable effort to schedule the treatment so as not to disrupt unduly the operations of the employer, subject to the approval of the health care provider of the employee or the health care provider of the son, daughter, spouse, or parent of the employee, as appropriate

b. Shall provide the employer with not less than 30 days' notice, before the date the leave is to begin, of the employee's intention to take leave under such subparagraph, except that if the date of the treatment requires leave to begin in less than 30 days, the employee shall provide such notice as is practicable.

Spouses Employed by the Same Employer

In any case in which a husband and wife entitled to leave under subsection (a) are employed by the same employer, the aggregate number of workweeks of leave to which both may be entitled may be limited to 12 workweeks during any 12-month period, if such leave is taken

1. Under subparagraph (A) or (B) of subsection (a)(1)
2. To care for a sick parent under subparagraph (C) of such subsection.

SEC. 103—CERTIFICATION

In General

An employer may require that a request for leave under subparagraph (C) or (D) of section 102(a)(1) be supported by a certification issued by the health care provider of the eligible employee or of the son, daughter, spouse, or parent of the employee, as appropriate. The employee shall provide, in a timely manner, a copy of such certification to the employer.

Sufficient Certification

Certification provided under subsection (a) shall be sufficient if it states

1. The date on which the serious health condition commenced
2. The probable duration of the condition
3. The appropriate medical facts within the knowledge of the health care provider regarding the condition

4. For purposes of leave
   a. Under section 102(a)(1)(C), a statement that the eligible employee is needed to care for the son, daughter, spouse, or parent and an estimate of the amount of time that such employee is needed to care for the son, daughter, spouse, or parent
   b. Under section 102(a)(1)(D), a statement that the employee is unable to perform the functions of the position of the employee
5. In the case of certification for intermittent leave, or leave on a reduced leave schedule, for planned medical treatment, the dates on which such treatment is expected to be given and the duration of such treatment
6. In the case of certification for intermittent leave, or leave on a reduced leave schedule, under section 102(a)(1)(D), a statement of the medical necessity for the intermittent leave or leave on a reduced leave schedule, and the expected duration of the intermittent leave or reduced leave e schedule
7. In the case of certification for intermittent leave, or leave on a reduced leave schedule, under section 102(a)(1)(C), a statement that the employee's intermittent leave or leave on a reduced leave schedule is necessary for the care of the son, daughter, parent, or spouse who has a serious health condition, or will assist in their recovery, and the expected duration and schedule of the intermittent leave or reduced leave schedule.

Second Opinion

1. In general—in any case in which the employer has reason to doubt the validity of the certification provided under subsection (a) for leave under subparagraph (C) or (D) of section 102(a)(1), the employer may require, at the expense of the employer, that the eligible employee obtain the opinion of a second health care provider designated or approved by the employer concerning any information certified under subsec-

tion (b) for such leave.

2. Limitation—a health care provider designated or approved under paragraph (1) shall not be employed on a regular basis by the employer.

Resolution of Conflicting Opinions

1. In general—in any case in which the second opinion described in subsection (c) differs from the opinion in the original certification provided under subsection (a), the employer may require, at the expense of the employer, that the employee obtain the opinion of a third health care provider designated or approved jointly by the employer and the employee concerning the information certified under subsection (b).

2. Finality—the opinion of the third health care provider concerning the information certified under subsection (b) shall be considered to be final and shall be binding on the employer and the employee.

Subsequent Rectification:
The employer may require that the eligible employee obtain subsequent recertification on a reasonable basis.

SEC. 104—EMPLOYMENT AND BENEFITS PROTECTION

Restoration to Position

1. In general—except as provided in subsection (b), any eligible employee who takes leave under section 102 for the intended purpose of the leave shall be entitled, on return from such leave
   a. To be restored by the employer to the position of employment held by the employee when the leave commenced
   b. To be restored to an equivalent position with equivalent employment benefits, pay, and other terms and conditions of employment.

2. Loss of benefits—the taking of leave under section 102 shall not result in the loss of any employment benefit accrued prior to the date on which the leave commenced

3. Limitations—nothing in this section shall be construed to entitle any restored employee to
   a. The accrual of any seniority or employment benefits during any period of leave
   b. Any right, benefit, or position of employment other than any right, benefit, or position to which the employee would have been enTitled had the employee not taken the leave

4. Certification—as a condition of restoration under paragraph (1) for an employee who has taken leave under section 102(a)(1)(D), the employer may have a uniformly applied practice or policy that requires each such employee to receive certification from the health care provider of the employee that the employee is able to resume work, except that nothing in this paragraph shall supersede a valid State or local law or a collective bargaining agreement that governs the return to work of such employees

5. Construction—nothing in this subsection shall be construed to prohibit an employer from requiring an employee on leave under section 102 to report periodically to the employer on the status and intention of the employee to return to work

Exemption for Certain Highly Compensated Employees

1. Denial of restoration—an employer may deny restoration under subsection (a) to any eligible employee described in paragraph (2) if:
   a. Such denial is necessary to prevent substantial and grievous economic injury to the operations of the employer
   b. The employer notifies the employee of the intent of the employer to deny restoration on such basis at the time the employer determines that such injury would occur

c. In any case in which the leave has commenced, the employee elects not to return to employment after receiving such notice.

2. Affected employees—an eligible employee described in paragraph (1) is a salaried eligible employee who is among the highest paid 10 percent of the employees employed by the employer within 75 miles of the facility at which the employee is employed.

Maintenance of Health Benefits

1. Coverage—except as provided in paragraph (2), during any period that an eligible employee takes leave under section 102, the employer shall maintain coverage under any "group health plan" (as defined in section 5000(b)(1) of the Internal Revenue Code of 1986) for the duration of such leave at the level and under the conditions coverage would have been provided if the employee had continued in employment continuously for the duration of such leave

2. Failure to return from leave—the employer may recover the premium that the employer paid for maintaining coverage for the employee under such group health plan during any period of unpaid leave under section 102 if
   a. The employee fails to return from leave under section 102 after the period of leave to which the employee is enTitled has expired
   b. The employee fails to return to work for a reason other than the continuation, recurrence, or onset of a serious health condition that enTitles the employee to leave under subparagraph (C) or (D) of section 102(a)(1) or other circumstances beyond the control of the employee

3. Certification
   a. Issuance—an employer may require that a claim that an employee is unable to return to work because of the continuation, recurrence, or onset of the serious health condition

described in paragraph (2)(B)(i) be supported by a certification issued by the health care provider of the son, daughter, spouse, or parent of the employee, as appropriate, in the case of an employee unable to return to work because of a condition specified in section 102(a)(1)(C) or a certification issued by the health care provider of the eligible employee, in the case of an employee unable to return to work because of a condition specified in section 102(a)(1)(D)

b. Copy—the employee shall provide, in a timely manner, a copy of such certification to the employer

c. Sufficiency of certification—leave because of serious health condition of employee or family member. The certification described in subparagraphs (A)(ii) and (A)(i) shall be sufficient if the certification states that a serious health condition prevented the employee from being able to perform the functions of the position of the employee or is needed to care for the son, daughter, spouse, or parent who has a serious health condition on the date that the leave of the employee expired.

SEC. 105—PROHIBITED ACTS

Interference with Rights

1. Exercise of rights—unlawful for any employer to interfere with, restrain, or deny the exercise of or the attempt to exercise, any right provided under this Title

2. Discrimination—unlawful for any employer to discharge or in any other manner discriminate against any individual for opposing any practice made unlawful by this Title

Interference with Proceedings or Inquiries

It shall be unlawful for any person to discharge or in any other manner discriminate against any individual because such individual

1. Has filed any charge, or has instituted or caused to be instituted any proceeding, under or related to this Title
2. Has given, or is about to give, any information in connection with any inquiry or proceeding relating to any right provided under this Title
3. Has testified, or is about to testify, in any inquiry or proceeding relating to any right provided under this Title

SEC. 106—INVESTIGATIVE AUTHORITY.

In General
To ensure compliance with the provisions of this Title, or any regulation or order issued under this Title, the Secretary shall have, subject to subsection (c), the investigative authority provided under section ll(a) of the Fair Labor Standards Act of 1938 (29 U.S.C. 211(a)).

Obligation to Keep and Preserve Records
Any employer shall make, keep, and preserve records pertaining to compliance with this Title in accordance with section 11(c) of the Fair Labor Standards Act of 1938 (29 U.S.C. 211(c)) and in accordance with regulations issued by the Secretary.

Required Submissions Limited to Annual Basis
The Secretary shall not under the authority of this section require any employer or any plan, fund, or program to submit to the Secretary any books or records more than once during any 12-month period, unless the Secretary has reasonable cause to believe there may exist a violation of this Title or any regulation or order issued pursuant to this Title, or is investigating a charge pursuant to section 107(b).

Subpoena Powers
For the purposes of any investigation provided for in this section, the Secretary shall have the subpoena authority provided for under section 9 of the Fair Labor Standards Act of 1938 (29 U.S.C. 209).

SEC. 107—ENFORCEMENT

Civil Actions by Employees

1. Liability—any employer who violates section 105 shall be liable to any eligible employee affected
   a. For damages equal to the amount of any wages, salary, employment benefits, or other compensation denied or lost to such employee by reason of the violation, or in a case in which wages, salary, employment benefits, or other compensation have not been denied or lost to the employee, any actual monetary losses sustained by the employee as a direct result of the violation, such as the cost of providing care, up to a sum equal to 12 weeks of wages or salary for the employee the interest on the amount described in clause (i) to be calculated at the prevailing rate, and an additional amount as liquidated damages equal to the sum of the amount described in clause (i) and the interest described in clause (ii), except that if an employer who has violated section 105 proves to the satisfaction of the court that the act or omission that violated section 105 was in good faith and that the employer had reasonable grounds for believing that the act or omission was not a violation of section 105, such court may, in the discretion of the court, reduce the amount of the liability to the amount and interest determined under clauses (i) and (ii), respectively
   b. For such equitable relief as may be appropriate, including employment, reinstatement, and promotion
2. Right of action—an action to recover the damages or equitable relief prescribed in paragraph (1) may be maintained against any employer (including a public agency) in any Federal or State court of competent jurisdiction by any one or more employees for and in behalf of
   a. The employees
   b. The employees and other employees similarly situated

3. Fees and costs—the court in such an action shall, in addition to any judgment awarded to the plaintiff, allow a reasonable attorney's fee, reasonable expert witness fees, and other costs of the action to be paid by the defendant.
4. Limitations—the right provided by paragraph (2) to bring an action by or on behalf of any employee shall terminate
   a. On the filing of a complaint by the Secretary in an action under subsection (d) in which restraint is sought of any further delay in the payment of the amount described in paragraph (1)(A) to such employee by an employer responsible under paragraph (1) for the payment
   b. On the filing of a complaint by the Secretary in an action under subsection (b) in which a recovery is sought of the damages described in paragraph (1)(A) owing to an eligible employee by an employer liable under paragraph (1) unless the action described in subparagraph (A) or (B) is dismissed without prejudice on motion of the Secretary

Action by the Secretary

1. Administrative action—the Secretary shall receive, investigate, and attempt to resolve complaints of violations of section 106 in the same manner that the Secretary receives, investigates, and attempts to resolve complaints of violations of sections 6 and 7 of the Fair Labor Standards Act of 1938 (29 U.S.C. 206 and 207).
2. Civil action—the Secretary may bring an action in any court of competent jurisdiction to recover the damages described in subsection (a)(1)(A).
3. Sums recovered—any sums recovered by the Secretary pursuant to paragraph (2) shall be held in a special deposit account and shall be paid, on order of the Secretary, directly to each employee affected. Any such sums not paid to an employee because of inability to do so within a period of 3 years shall be deposited into the Treasury of the United States as miscellaneous receipts.

Limitation

1. In general—except as provided in paragraph (2), an action may be brought under this section not later than 2 years after the date of the last event constituting the alleged violation for which the action is brought.
2. Willful violation—in the case of such action brought for a willful violation of section 105, such action may be brought within 3 years of the date of the last event constituting the alleged violation for which such action is brought.
3. Commencement—in determining when an action is commenced by the Secretary under this section for the purposes of this subsection, it shall be considered to be commenced on the date when the complaint is filed.

Action for Injunction by Secretary

The district courts of the United States shall have jurisdiction, for cause shown, in an action brought by the Secretary

1. To restrain violations of section 105, including the restraint of any withholding of payment of wages, salary, employment benefits, or other compensation, plus interest, found by the court to be due to eligible employees
2. To award such other equitable relief as may be appropriate, including employment, reinstatement, and promotion

Solicitor of Labor

The Solicitor of Labor may appear for and represent the Secretary on any litigation brought under this section.

SEC. 108—SPECIAL RULES CONCERNING EMPLOYEE OF LOCAL EDUCATIONAL AGENCIES

Application

1. In general—except as otherwise provided in this section, the rights (including the rights under section 104, which shall

extend throughout the period of leave of any employee under this section), remedies, and procedures under this Title shall apply to

   a. Any "local educational agency" (as defined in section 1471(12) of the Elementary and Secondary Education Act of 1965 (20 U.S.C. 2891(12)) and an eligible employee of the agency

   b. Any private elementary or secondary school and an eligible employee of the school

2. Definitions—for purposes of the application described in paragraph

   a. Eligible employee—the term eligible employee means an eligible employee of an agency or school described in paragraph (1).

   b. Employer—the term "employer" means an agency or school described in paragraph (1).

Leave Does Not Violate Certain Other Federal Laws

A local educational agency and a private elementary or secondary school shall not be in violation of the Individuals with Disabilities Education Act (20 U.S.C. 1400 et seq.), section 504 of the Rehabilitation Act of 1973 (29 U.S.C. 794), or Title VI of the Civil Rights Act of 1964 (42 U.S.C. 2000d et seq.), solely as a result of an eligible employee of such agency or school exercising the rights of such employee under this Title.

Intermittent or Reduced Instructional Employees' Leave

1. In general—subject to paragraph (2), in any case in which an eligible employee employed principally in an instructional capacity by any such educational agency or school requests leave under subparagraph (C) or (D) of section 102(a)(1) that is foreseeable based on planned medical treatment and the employee would be on leave for greater than 20 percent of the total number of working days in the period during which the

leave would extend, the agency or school may require that such employee elect either

    a. To take leave for periods of a particular duration, not to exceed the duration of the planned medical treatment

    b. To transfer temporarily to an available alternative position offered by the employer for which the employee is qualified, has equivalent pay and benefits, and better accommodates recurring periods of leave than the regular employment position of the employee

2. application—the elections described in subparagraphs (A) and (B) of paragraph (1) shall apply only with respect to an eligible employee who complies with section 102(e)(2).

Rules Applicable to Periods Near the End of a Term

The following rules shall apply with respect to periods of leave near the conclusion of an academic term in the case of any eligible employee employed principally in an instructional capacity by any such educational agency or school:

1. Leave more than five weeks prior to end of term—if the eligible employee begins leave under section 102 more than 5 weeks prior to the end of the academic term, the agency or school may require the employee to continue taking leave until the end of such term

    a. The leave is of at least 3 weeks duration

    b. The return to employment would occur during the 3 week period before the end of such term

2. Leave less than five weeks prior to end of term—if the eligible employee begins leave under subparagraph (A), (B), or (C) of section 102(a)(1) during the period that commences 5 weeks prior to the end of the academic term, the agency or school may require the employee to continue taking leave until the end of such term, if

    a. the leave is of greater than two weeks duration

b. The return to employment would occur during the 2 week period before the end of such term

3. Leave less than three weeks prior to end of term—if the eligible employee begins leave under subparagraph (A), (B), or (C) of section 102(a)(1) during the period that commences 3 weeks prior to the end of the academic term and the duration of the leave is greater than 5 working days, the agency or school may require the employee to continue to take leave until the end of such term.

Restoration to Equivalent Employment Compensation

For purposes of determinations under section 104(a)(1)(B) (relating to the restoration of an eligible employee to an equivalent position), in the case of a local educational agency or a private elementary or secondary school, such determination shall be made on the basis of established school board policies and practices, private school policies and practices, and collective bargaining agreements.

Reduction of the Amount of Liability

If a local educational agency or a private elementary or secondary school that has violated this Title proves to the satisfaction of the court that the agency, school, or department had reasonable grounds for believing that the underlying act or omission was not a violation of this Title, such court may, in the discretion of the court, reduce the amount of the liability provided for under section 107(a)(1)(A) to the amount and interest determined under clauses (i) and (ii), respectively, of such section.

SEC. 109—NOTICE

In General

Each employer shall post and keep posted, in conspicuous places on the premises of the employer where notices to employees and applicants for employment are customarily posted, a notice, to be prepared or approved by the Secretary, setting forth excerpts from, or summaries of, the pertinent provisions of this Title and information pertaining to the filing of a charge.

Penalty
Any employer that willfully violates this section may be assessed a civil money penalty not to exceed $100 for each separate offense.

SEC. 401—EFFECT ON OTHER LAWS

Federal and State Antidiscrimination Laws
Nothing in this Act or any amendment made by this Act shall be construed to modify or affect any Federal or State law prohibiting discrimination on the basis of race, religion, color, national origin, sex, age, or disability.

State and Local Laws
Nothing in this Act or any amendment made by this Act shall be construed to supersede any provision of any State or local law that provides greater family or medical leave rights than the rights established under this Act or any amendment made by this Act.

SEC. 402—EFFECT ON EXISTING EMPLOYMENT BENEFITS

More Protective
Nothing in this Act or any amendment made by this Act shall be construed to diminish the obligation of an employer to comply with any collective bargaining agreement or any employment benefit program or plan that provides greater family or medical leave rights to employees than the rights established under this Act or any amendment made by this Act.

Less Protective
The rights established for employees under this Act or any amendment made by this Act shall not be diminished by any collective bargaining agreement or any employment benefit program or plan.

SEC. 403 — ENCOURAGEMENT OF MORE GENEROUS LEAVE POLICIES.

Nothing in this Act or any amendment made by this Act shall be construed to discourage employers from adopting or retaining leave policies more generous than any policies that comply with the requirements under this Act or any amendment made by this Act.

SEC. 404 — REGULATIONS.

The Secretary of Labor shall prescribe such regulations as are necessary to carry out Title I and this Title not later than 120 days after the date of the enactment of this Act.

SEC. 406 — EFFECTIVE DATES.

Title III:
Title III shall take effect on the date of the enactment of this Act.

OTHER TITLES:

1. In general — except as provided in paragraph (2), Titles I, II, and V and this Title shall take effect 6 months after the date of the enactment of this Act.
2. Collective bargaining agreements — in the case of a collective bargaining agreement in effect on the effective date prescribed by paragraph (1), Title I shall apply on the earlier of
   a. The date of the termination of such agreement; or
   b. The date that occurs 12 months after the date of the enactment of this Act.

✦ ✦ ✦

Fragante v. City & County of Honolulu
888 F.2d 591 (9th Cir.1989), cert. denied,
494 U.S. 1081 (1990).

Manuel Fragante applied for a clerk's job with the City and County of Honolulu (Defendants). Although he placed high enough on a civil service eligible list to be chosen for the position, he was not selected because of a perceived deficiency in relevant oral communication skills caused by his "heavy Filipino accent." Fragante brought suit, alleging that the defendants discriminated against him on the basis of his national origin, in violation of Title VII of the Civil Rights Act. At the conclusion of a trial, the district court found that the oral ability to communicate effectively and clearly was a legitimate occupational qualification for the job in question. This finding was based on the court's understanding that an important aspect of defendant's business, for which a clerk would be responsible, involved the providing of services and assistance to the general public. The court also found that defendant's failure to hire Fragante was explained by his deficiencies in the area of oral communication, not because of his national origin. Finding no proof of a discriminatory intent or motive by the defendant the court dismissed Fragante's complaint, and he appeals. We have jurisdiction under 28 U.S.C. § 1291, and we affirm.

Facts

In April 1981, at the age of sixty, Fragante emigrated from the Philippines to Hawaii. In response to a newspaper ad, he applied in November of 1981 for the job at issue in this appeal—an entry level Civil Service Clerk SR-8 job for the City of Honolulu's Division of Motor Vehicles and Licensing. The SR-8 clerk position involved such tasks as filing, processing mail, cashiering, orally providing routine information to the "sometimes contentious" public over the telephone and at an information counter, and obtaining supplies. Fragante scored the highest of 721 test takers on the written SR-8 Civil Service Examination that tested, among other things, word usage, grammar and spelling. Accordingly, he was ranked first on a certified list of eligible for two SR-8 clerk positions, an achievement of which he is understandably quite proud.

Fragante then was interviewed in the normal course of the selection process, as were other applicants, by George Kuwahara, the assistant licensing administrator, and Kalani McCandless, the division secretary. Both Kuwahara and McCandless were personally familiar with the demands of the position at issue, and both had extensive experience interviewing applicants to the division. During the interview, Kuwahara stressed that the position involved constant public contact and that the ability to speak clearly was one of the most important skills required for the position.

Both Kuwahara and McCandless had difficulty understanding Fragante because of his pronounced Filipino accent, and they determined on the basis of the oral interview that he would be difficult to understand both at the information counter and over the telephone. Accordingly, both interviewers gave Fragante a negative recommendation. They noted he had a very pronounced accent and was difficult to understand. It was their judgment that this would interfere with his performance of certain aspects of the job. As a consequence, Mr. Fragante dropped from number one to number three on the list of eligible for the position.

Under the city's civil service rules, the Department of Motor Vehicles and Licensing, as the appointing authority, is allowed discretion in selecting applicants for the clerk vacancies. City Civil Service Rule 4.2(d) allows the defendants to select any of the top five eligible without regard to their rank order. The essence of this rule was clearly stated in the employment announcement posted for the SR-8 position:

The names of the "top five" qualified applicants with the highest examination grades will be referred to the employing agency in the order of their examination grade and availability for employment according to Civil Service Rules. The employing agency may select any one of the eligible referred. Those not selected will remain on the list for at least one year for future referrals.

In accord with this process, the two other applicants who were judged more qualified than Fragante and who therefore placed higher than he on the final list got the two available jobs, and he was so notified by mail.

After exhausting administrative remedies, Fragante filed a claim under Title VII of the Civil Rights Act against the City and County of Honolulu, alleging he was discriminated against because of his accent. The district court relied on the results of the oral interview and found that Fragante's oral skills were "hampered by his accent or manner of speaking." The court found no evidence of unlawful discrimination in violation of Title VII, concluding that Fragante lacked the "occupational requirement" of being able to communicate effectively with the public, and dismissed his claim.

Discussion

Title VII prohibits employment discrimination on the basis of race, color, sex, religion and national origin. 42 U.S.C. § 2000e-2(a)(1) (1982). A plaintiff may bring an action against an employer under a disparate treatment and/or disparate impact theory. Fragante's action was brought under the disparate treatment theory.

1. Prima Facie Case

Defendants first argue Fragante failed to meet his burden of proving a *prima facie* case because he failed to show he was actually qualified for the SR-8 clerk position, a position that requires the applicant to be able to communicate clearly and effectively. Fragante, on the other hand, contends he was qualified for the position. As proof he points to his exceptional score on the objective written examination, and he argues that his speech, though heavily accented, was deemed comprehensible by two expert witnesses at trial. Fragante's position is supported by the approach taken by the Equal Employment Opportunity Commission which submits that a plaintiff who proves he has been discriminated against solely because of his accent does establish a *prima facie* case of national origin discrimination. This contention is further supported by EEOC guidelines that define discrimination to include "the denial of equal employment opportunity... because an individual has the...linguistic characteristics of a national origin group." 29 C.F.R. § 1606.1 (1988). Furthermore, Fragante was never advised that he was not qualified for the job; he was only told that he was less qualified than his competition.

Because we find that Fragante did not carry the ultimate burden of proving national origin discrimination, however, the issue of whether Fragante established a *prima facie* case of discrimination is not significant, and we assume without deciding that he did.

2. The Statute and its Purpose

Preliminarily, we do well to remember that this country was founded and has been built in large measure by people from other lands, many of whom came here, especially after our beginnings, with a limited knowledge of English. This flow of immigrants has continued and has been encouraged over the years. From its inception, the United States of America has been a dream to many people around the world. We hold out promises of freedom, equality, and economic opportunity to many who only know these words as concepts. It would be more than ironic if we followed up our invitation to people such as Manuel Fragante with a closed economic door based on national origin discrimination. It is no surprise that Title VII speaks to this issue and clearly articulates the policy of our nation: unlawful discrimination based on national origin shall not be permitted to exist in the workplace. But, it is also true that there is another important aspect of Title VII: the "preservation of an employer's remaining freedom of choice."

3. Proof of an Ultimate Case of Discrimination

We turn our discussion to whether defendants articulated a legitimate, nondiscriminatory reason for Fragante's nonselection. We find that they did, but to this finding we add a note of caution to the trial courts. Accent and national origin are obviously inextricably intertwined in many cases. It would therefore be an easy refuge in this context for an employer unlawfully discriminating against someone based on national origin to state falsely that it was not the person's national origin that caused the employment or promotion problem, but the candidate's inability to measure up to the communications skills demanded by the job. We encourage a very searching look by the district courts at such a claim.

An adverse employment decision may be predicated upon an individual's accent when, but only when, it interferes materially with job performance. There is nothing improper about an employer making an *honest* assessment of the oral communications skills of a candidate for a job when such skills are reasonably related to job performance.

The defendants advertised for applicants to fill SR-8 vacancies. The initial job announcement listed the ability to "deal tactfully and effectively with the public" as one of the areas to be tested. There is no doubt from the record that the oral ability to communicate effectively in English is reasonably related to the normal operation of the clerk's office. A clerk must be able to respond to the public's questions in a manner that the public can understand. In this regard, the district court in its Findings of Fact and Conclusions of Law and Order made the following significant observations:

1. The job is a difficult one because it involves dealing with a great number of disgruntled members of the public.
2. The clerk must deal with 200–300 people a day, many of whom are angry or complaining and who do not want to hear what the clerk may have to explain concerning their applications or an answer to their questions.
3. It is a high turnover position where people leave quickly because of the high stress involving daily contact with contentious people.

What must next be determined is whether defendants established a factual basis for believing that Fragante would be hampered in performing this requirement. Defendants submit that because his accent made Fragante difficult to understand as determined by the interview, he would be less able to perform the job than other applicants. Fragante, on the other hand, contends he is able to communicate effectively in English as established by two expert witnesses at trial and by his responses in open court. In essence, he argues his non-selection was effectively based upon national origin discrimination.

After the interview, Kuwahara and McCandless scored Fragante on a rating sheet that was used for all applicants. Applicants were scored in the categories of appearance, speech, self-confidence, emotional control, alertness, initiative, personality, attitude, work experience, and overall fitness for the job. A scale of 1–10 was used. Kuwahara gave Fragante a score of 3 for speech, and noted: "...very pronounced accent, difficult to understand." Although McCandless did not enter a score in the speech category, she noted: "...heavy Filipino accent; would be difficult to understand over the telephone."

After the interviews were scored, Kuwahara and McCandless reviewed the scores, discussed the applicants, and decided on their hiring recommendation to finance director Peter Leong. In making the recommendation, written examination scores were given no consideration. Kuwahara prepared the written recommendation to Leong, dated April 13, 1982, recommending two others for selection. Fragante in his position as Number 3 on the final list was described as follows:

Manuel Fragante—Retired Phillippine (*sic*) army officer. Speaks with very pronounced accent that is difficult to understand. He has 37 years of experience in management administration and appears more qualified for professional rather than clerical work. However, because of his accent, I would not recommend him for this position.

McCandless then notified Fragante that he was not selected for either of the clerk position vacancies. Pursuant to a request from

Fragante, Kuwahara then reduced the matter to writing. In a letter, dated June 28, 1982, the reasons why he was not selected were articulated as follows:

"As to the reason for your non-selection, we felt the two selected applicants were both superior in their verbal communication ability. As we indicated in your interview, our clerks are constantly dealing with the public and the ability to speak clearly is one of the most important skills required for the position. Therefore, while we were impressed with your educational and employment history, we felt the applicants selected would be better able to work in our office because of their communication skills."

Thus, the interviewers' record discloses Fragante's third place ranking was based on his "pronounced accent that is difficult to understand." Indeed, Fragante can point to no facts that indicate that his ranking was based on factors other than his inability to communicate effectively with the public. This view was shared by the district court.

Although the district court determined that the interview lacked some formality as to standards, instructions, guidelines, or criteria for its conduct and that the rating sheet was inadequate, the court also found that these "insufficiencies" were irrelevant with respect to plaintiff's complaint of unlawful discrimination. A review of the record reveals nothing that would impeach this assessment. Kuwahara and McCandless recorded their evaluation of Fragante's problem in separate written remarks on their rating sheets. As such, they established a legitimate factual basis for the conclusion that Fragante would be less able than his competitors to perform the required duties.

Fragante argues the district court erred in considering "listener prejudice" as a legitimate, nondiscriminatory reason for failure to hire. We find, however, that the district court did not determine defendants refused to hire Fragante on the basis that some listeners would "turn off" a Filipino accent. The district court after trial noted that: "Fragante, in fact, has a difficult manner of pronunciation and the Court further finds as a fact from his general testimony that he would often not respond directly to the questions as propounded. He maintains much of his military bearing." We regard the last sentence of the court's comment to be little more than a stray remark of no moment.

We do not find the court's conclusion clearly erroneous. We find support for our view in *Fernandez v. Wynn Oil.*, 653 F.2d 1273, 1275 (9th Cir.1981), where this court held inability to communicate effectively to be one valid ground for finding a job applicant not qualified.

Having established that defendants articulated a legitimate reason for Fragante's non-selection, our next inquiry is whether the reason was a mere pretext for discrimination. Fragante essentially argues that defendant's selection and evaluation procedures were so deficient as to render the proffered reason for non-selection nothing more than a pretext for national origin discrimination. The problem with this argument, however, is that on examination it is only a charge without substance. The process may not have been perfect, but it reveals no discriminatory motive or intent. Search as we have, we have not been able to find even a hint of a mixed motive such as existed in *Price Waterhouse*. Instead, it appears that defendants were motivated exclusively by reasonable business necessity.

Fragante's counsel attempts to cast this case as one in which his client was denied a job simply because he had a difficult accent. This materially alters what actually happened. Fragante failed to get the job because two competitors had superior qualifications with respect to a relevant task performed by a government clerk. Insofar as this implicates "the interest of the State, as an employer, in promoting the efficiency of the public services it performs through its employees ... it is not something we are permitted to ignore." Title VII does not stand for the proposition that a person in a protected class, or a person with a foreign accent, shall enjoy a position of advantage thereby when competing for a job against others not similarly protected. And, the record does not show that the jobs went to persons less qualified than Fragante: to the contrary.

Under our holding in *Ward v. Westland Plastics, Inc.*, 651 F. 1266, 1269 (9th Cir.1980), "[a]n employer's decision may be justified by the hired employee's superior qualifications unless the purported justification is a pretext for invidious discrimination." In this case, there is simply no proof whatsoever of pretext, and we do not find the district court's finding of "no discrimination" to be clearly erroneous.

In sum, the record conclusively shows that Fragante was passed over because of the deleterious *effect* of his Filipino accent on his ability to communicate orally, not merely because he had such an accent.

The district court is AFFIRMED.

EMPLOYEE POLYGRAPH PROTECTION ACT

§ 2001. Definitions

As used in this chapter:

1. Commerce—the meaning provided by section 203(b) of this Title.
2. Employer—any person acting directly or indirectly in the interest of an employer in relation to an employee or prospective employee.
3. Lie detector—includes a polygraph, deceptograph, voice stress analyzer, psychological stress evaluator, or any other similar device (whether mechanical or electrical) that is used, or the results of which are used, for the purpose of rendering a diagnostic opinion regarding the honesty or dishonesty of an individual.
4. Polygraph—an instrument that:
   a. Records continuously, visually, permanently, and simultaneously changes in cardiovascular, respiratory, and electrodermal patterns as minimum instrumentation standards
   b. Is used, or the results of which are used, for the purpose of rendering a diagnostic opinion regarding the honesty or dishonesty of an individual.
5. Secretary—the Secretary of Labor

§ 2002. Prohibitions on lie detector use

Except as provided in sections 2006 and 2007 of this Title, it shall be unlawful for any employer engaged in or affecting commerce or in the production of goods for commerce:

1. Directly or indirectly, to require, request, suggest, or cause any employee or prospective employee to take or submit to any lie detector test
2. To use, accept, refer to, or inquire concerning the results of any lie detector test of any employee or prospective employee
3. To discharge, discipline, discriminate against in any manner, or deny employment or promotion to, or threaten to take any such action against:
   a. Any employee or prospective employee who refuses, declines, or fails to take or submit to any lie detector test
   b. Any employee or prospective employee on the basis of the results of any lie detector test
4. To discharge, discipline, discriminate against in any manner, or deny employment or promotion to, or threaten to take any such action against, any employee or prospective employee because:
   a. Such employee or prospective employee has filed any complaint or instituted or caused to be instituted any proceeding under or related to this chapter,
   b. Such employee or prospective employee has testified or is about to testify in any such proceeding, or
   c. Of the exercise by such employee or prospective employee, on behalf of such employee or another person, of any right afforded by this chapter.

§ 2003. Notice of protection

The Secretary shall prepare, have printed, and distribute a notice setting forth excerpts from, or summaries of, the pertinent provisions of this chapter. Each employer shall post and maintain such notice in conspicuous places on its premises where notices to employees and applicants to employment are customarily posted.

§ 2004. Authority of the Secretary

In general

The Secretary shall:

1. Issue such rules and regulations as may be necessary or appropriate to carry out this chapter
2. Cooperate with regional, State, local, and other agencies, and cooperate with and furnish technical assistance to employers, labor organizations and employment agencies to aid in effectuating the purposes of this chapter
3. Make investigations and inspections and require the keeping of records necessary or appropriate for the administration of this chapter

SUBPOENA AUTHORITY

For the purpose of any hearing or investigation under this chapter, the Secretary shall have the authority contained in sections 49 and 50 of Title 15.

§ 2005. Enforcement provisions

Civil penalties

1. In general—subject to paragraph (2), any employer who violates any provision of this chapter may be assessed a civil penalty of not more than $10,000.
2. Determination of amount—in determining the amount of any penalty under paragraph (1), the Secretary shall take into

account the previous record of the person in terms of compliance, with this chapter and the gravity of the violation.

3. Collection—any civil penalty assessed under this subsection shall be collected in the same manner as is required by subsections (b) through (e) of section 1853 of this Title with respect to civil penalties assessed under subsection (a) of such section.

Injunctive actions by the Secretary

The Secretary may bring an action under this section to restrain violations of this chapter. The Solicitor of Labor may appear for and represent the Secretary in any litigation brought under this chapter. In any action brought under this section, the district courts of the United States shall have jurisdiction, for cause shown, to issue temporary or permanent restraining orders and injunctions to require compliance with this chapter, including such legal or equitable relief incident thereto as may be appropriate, including, but not limited to, employment, reinstatement, promotion, and the payment of lost wages and benefits.

Private civil actions

1. Liability—an employer who violated this chapter shall be liable to the employee or prospective employee affected by such violation. Such employer shall be liable for such legal or equitable relief as may be appropriate, including, but not limited to, employment, reinstatement, promotion, and the payment of lost wages and benefits

2. Court—an action to recover the liability prescribed in paragraph (1) may be maintained against the employer in any Federal or State court of competent jurisdiction by an employee or prospective employee for or on behalf of such employee, prospective employee, and other employees or prospective employees similarly situated. No such action may be commenced more than 3 years after the date of the alleged violation

3. Costs—the court, in its discretion, may allow the prevailing party (other than the United States) reasonable costs, including attorney's fees

Waiver of rights prohibited
The rights and procedures provided by this chapter may not be waived by contract or otherwise, unless such waiver is part of a written settlement agreed to and signed by the parties to the pending action or complaint under this chapter.

§ 2006. Exemptions

No application to Governmental employers
This chapter shall not apply with respect to the United States Government, any State or local government, or any political subdivision of a State or local government.

National defense and security exemption

1. National defense—nothing in this chapter shall be construed to prohibit the administration, by the Federal Government, in the performance of any counterintelligence function, of any lie detector test to:
   a. Any expert or consultant under contract to the Department of Defense or any employee of any contractor of such Department
   b. Any expert or consultant under contract with the Department of Energy in connection with the atomic energy defense activities of such Department or any employee of any contractor of such Department in connection with such activities.
2. Security—nothing in this chapter shall be construed to prohibit the administration, by the Federal Government, in the performance of any intelligence or counterintelligence function, of any lie detector test to:

a. Any individual employed by, assigned to, or detailed to, the National Security Agency, the Defense Intelligence Agency, or the Central Intelligence Agency,

b. Any expert or consultant under contract to any such agency,

c. Any employee of a contractor to any such agency,

d. Any individual applying for a position in any such agency, or

e. Any individual assigned to a space where sensitive cryptologic information is produced, processed, or stored for any such agency; or

f. Any expert, or consultant (or employee of such expert or consultant) under contract with any Federal Government department, agency, or program whose duties involve access to information that has been classified at the level of top secret or designed as being within a special access program under the section 4.2(a) of Executive Order 12356 (or a successor Executive order).

FBI contractors exemption

Nothing in this chapter shall be construed to prohibit the administration, by the Federal Government, in the performance of any counterintelligence function, of any lie detector test to an employee of a contractor of the Federal Bureau of Investigation of the Department of Justice who is engaged in the performance of any work under the contract with such Bureau.

Limited exemption for ongoing investigations

Subject to sections 2007 and 2009 of this Title, this chapter shall not prohibit an employer from requesting an employee to submit to a polygraph test if:

1. The test is administered in connection with an ongoing investigation involving economic loss or injury to the employer's business, such as theft, embezzlement, misappropriation, or an act of unlawful industrial espionage or sabotage;

2. The employee had access to the property that is the subject of the investigation;
3. The employer has a reasonable suspicion that the employee was involved in the incident or activity under investigation; and the employer executes a statement, provided to the examinee before the test, that:
   a. Sets forth with particularity the specific incident or activity being investigated and the basis for testing particular employees,
   b. Is signed by a person (other than a polygraph examiner) authorized to legally bind the employer
   c. Is retained by the employer for at least 3 years
   d. Contains at a minimum:
      • An identification of the specific economic loss or injury to the business of the employer
      • A statement indicating that the employee had access to the property that is the subject of the investigation
      • A statement describing the basis of the employer's reasonable suspicion that the employee was involved in the incident or activity under investigation

Exemption for security services

1. In general—subject to paragraph (2) and sections 2007 and 2009 of this Title, this chapter shall not prohibit the use of polygraph tests on prospective employees by any private employer whose primary business purpose consists of providing armored car personnel, personnel engaged in the design, installation, and maintenance of security alarm systems, or other uniformed or plainclothes security personnel and whose function includes protection of:
   a. Facilities, materials, or operations having a significant impact on the health or safety of any State or political subdivision thereof, or the national security of the United States, as determined under rules and regulations issued by the Secretary within 90 days after June 27, 1988, including:

- Facilities engaged in the production, transmission, or distribution of electric or nuclear power,
- Public water supply facilities,
- Shipments or storage of radioactive or other toxic waste materials
- Public transportation,

b. Currency, negotiable securities, precious commodities or instruments, or proprietary information.

2. Access—the exemption provided under this subsection shall not apply if the test is administered to a prospective employee who would not be employed to protect facilities, materials, operations, or assets referred to in paragraph (1).

Exemption for drug security, drug theft, or drug diversion investigations

1. In general—subject to paragraph (2) and sections 2007 and 2009 of this Title, this chapter shall not prohibit the use of a polygraph test by any employer authorized to manufacture, distribute, or dispense a controlled substance listed in schedule I, II, III, or IV of section 812 of Title 21.

2. Access—the exemption provided under this subsection shall apply:

a. If the test is administered to a prospective employee who would have direct access to the manufacture, storage, distribution, or sale of any such controlled substance

b. In the case of a test administered to a current employee, if:

- The test is administered in connection with an ongoing investigation of criminal or other misconduct involving, or potentially involving, loss or injury to the manufacture, distribution, or dispensing of any such controlled substance by such employer

- The employee had access to the person or property that is the subject of the investigation

§ 2007. Restrictions on use of exemptions

Test as basis for adverse employment action

1. Under ongoing investigations exemption—except as provided in paragraph (2), the exemption under subsection (d) of section 2006 of this Title shall not apply if an employee is discharged, disciplined, denied employment or promotion, or otherwise discriminated against in any manner on the basis of the analysis of a polygraph test chart or the refusal to take a polygraph test, without additional supporting evidence. The evidence required by such subsection may serve as additional supporting evidence.
2. Under other exemptions—in the case of an exemption described in subsection (e) or (f) of such section, the exemption shall not apply if the results of an analysis of a polygraph test chart are used, or the refusal to take a polygraph test is used, as the sole basis upon which an adverse employment action described in paragraph (1) is taken against an employee or prospective employee.

Rights of examinee

The exemptions provided under subsections (d), (e), and (f) of section 2006 of this Title shall not apply unless the requirements described in the following paragraphs are met:

1. All phases—throughout all phases of the test:
   a. The examinee shall be permitted to terminate the test at any time
   b. The examinee is not asked questions in a manner designed to degrade, or needlessly intrude on, such examinee
   c. the examinee is not asked any question concerning:
      • Religious beliefs or affiliations,
      • Beliefs or opinions regarding racial matters,
      • Political beliefs or affiliations,

- Any matter relating to sexual behavior
- Beliefs, affiliations, opinions, or lawful activities regarding unions or labor organizations

d. The examiner does not conduct the test if there is sufficient written evidence by a physician that the examinee is suffering from a medical or psychological condition or undergoing treatment that might cause abnormal responses during the actual testing phase

2. Pretest phase—during the pretest phase, the prospective examinee:

a. Is provided with reasonable written notice of the date, time, and location of the test, and of such examinee's right to obtain and consult with legal counsel or an employee representative before each phase of the test

b. Is informed in writing of the nature and characteristics of the tests and of the instruments involved

c. Is informed, in writing:
- Whether the testing area contains a two-way mirror, a camera, or any other device through which the test can be observed
- Whether any other device, including any device for recording or monitoring the test, will be used
- That the employer or the examinee may (with mutual knowledge) make a recording of the test

d. Is read and signs a written notice informing such examinee:
- That the examinee cannot be required to take the test as a condition of employment
- That any statement made during the test may constitute additional supporting evidence for the purposes of an adverse employment action described in subsection (a) of this section
- Of the limitations imposed under this section
- Of the legal rights and remedies available to the examinee if the polygraph test is not conducted in accordance with this chapter

- Of the legal rights and remedies of the employer under this chapter (including the rights of the employer under section 2008(c)(2) of this Title)
  e. Is provided an opportunity to review all questions to be asked during the test and is informed of the right to terminate the test at any time.
3. Actual testing phase—During the actual testing phase, the examiner does not ask such examinee any question relevant during the test that was not presented in writing for review to such examinee before the test.
4. Protection phase—before any adverse employment action, the employer shall:
   a. Further interview the examinee on the basis of the results of the test; and
   b. Provide the examinee with:
      - (i) A written copy of any opinion or conclusion rendered as a result of the test
      - (ii) A copy of the questions asked during the test along with the corresponding charted responses
5. Maximum number and minimum duration of tests—the examiner shall not conduct and complete more than five polygraph tests on a calendar day on which the test is given, and shall not conduct any such test for less than a 90-minute duration.

Qualifications and requirement of examiners
The exemptions provided under subsections (d), (e), and (f) of section 2006 of this Title shall not apply unless the individual who conducts the polygraph test satisfies the requirements under the following paragraphs:
   (1) Qualifications—the examiner:
      a. Has a valid and current license granted by licensing and regulatory authorities in the State in which the test is to be conducted, if so required by the State
      b. Maintains a minimum of a $50,000 bond or an equivalent amount of professional liability coverage.

(2) Requirements—the examiner:

   a. Renders any opinion or conclusion regarding the test
- In writing and solely on the basis of an analysis of polygraph test charts
- That does not contain information other than admissions, information, case facts and interpretation of the charts relevant to the purpose and stated objectives of the test
- That does not include any recommendation concerning the employment of the examinee

   b. Maintains all opinions, reports, charts, written questions, lists, and other records relating to the test for a minimum period of 3 years after administration of the test.

§ 2008. Disclosure of information

In general

A person, other than the examinee, may not disclose information obtained during a polygraph test, except as provided in this section.

Permitted disclosures

A polygraph examiner may disclose information acquired from a polygraph test only to:

1. the examinee or any other person specifically designed in writing by the examinee
2. the employer that requested the test
3. any court, governmental agency, arbitrator, or mediator, in accordance with due process of law, pursuant to an order from a court of competent jurisdiction.

Disclosure by employer

An employer (other than an employer described in subsection (a), (b), or (c) of section 2006 of this Title) for whom a polygraph test is conducted may disclose information from the test only to:

1. A person in accordance with subsection (b) of this section
2. A governmental agency, but only insofar as the disclosed information is an admission of criminal conduct.

§ 2009. Effect on other laws and agreements

Except as provided in subsections (a), (b), and (c) of section 2006 of this Title, this chapter shall not preempt any provision of any State or local law or of any negotiated collective bargaining agreement that prohibits lie detector tests or is more restrictive with respect to lie detector tests than any provision of this chapter.

10

Liabilities in Termination of Employment

Things could be worse. Suppose your errors were counted and published every day, like those of a baseball player.

Anon.

Every year, thousands of Americans are voluntarily or involuntarily terminated from their employment in the fire service and with other employers because of a multitude of reasons ranging from their failure at adhering to organization policies to the company's "downsizing" for economic purposes. In general terms, there are three basic ways in someone leaves a job: (1) voluntary termination or the employee simply quits the job, (2) involuntary termination or the person is "fired" from the job, and (3) a constructive discharge where the organization requests or forces the employee to resign from the job. In the past, an individual who was involuntarily terminated or constructively discharged from the job had few, if any, recourse against the company or organization for the termination.

In the United States, the traditional common law rule of employment at-will is the basis for which most employment situations are analyzed. The common law rule of employment at-will, in essence, means that an employee can be terminated from employment at any time, without notice and with or without cause by the employer.[1] However, if the employee has benefit of a contractual agreement, like a collective bargaining agreement between a labor organization and company, or is protected by federal or state statute, the contractual relationship or law removes the employee from the at-will status.

Employment at-will is a well established doctrine that has been recognized in the United States since the late 1800s. As stated in the *Payne v. Western & Atlantic Railroad*[2] case in 1884:

All (employers) may dismiss their employees at will, be they many or few, for good cause, for no cause, or even for cause normally wrong, without being thereby guilty of legal wrong.

WRONGFUL DISCHARGE

Although the employment at-will doctrine continues to be well entrenched in the American judicial system, numerous state and federal statutes have greatly eroded the right of an employer to terminate an individual's employment without cause. Some of the federal laws that have limited the employment at-will doctrine by prohibiting retaliation or discrimination against the employee include:

1. Labor Management Relations Act (includes protections for employees to form or join a union)[3]
2. Fair Labor Standards Act (includes protection for individuals who report violations)[4]
3. Title VII, Civil Rights Act of 1964 (discrimination to a protected class)[5]
4. Age Discrimination in Employment Act of 1967 (discrimination against employees over the age of 40)[6]

---

[1] H.G. Wood, *Law of Master and Servant*, 2d Ed., § 134 (NY: J.D. Parsons, Jr. 1877, 272–273.

[2] 81 Tenn. 507 (1884).

[3] 29 U.S.C.A. § 151, et.seq.

[4] 29 U.S.C.A. § 215.

[5] 42 U.S.C.A. § 2000e-2 and § 2000e-3(a).

[6] 29 U.S.C.A. § 621 & 734.

5. Rehabilitation Act of 1973 (disability protection)[7]
6. Americans With Disabilities Act (disability protection)[8]
7. Employee Retirement Income Security Act of 1974 ("ERISA" —protection for employees close to retirement)[9]
8. The Vietnam Era Veterans Readjustment Assistance Act (protections to Vietnam veterans)[10]
9. Occupational Safety and Health Act (protection for employees reporting violations)[11]
10. Clean Air Act (protection for employees reporting violations)[12]
11. Civil Service Reform Act of 1978 (special protections for individuals in the civil service system)[13]
12. Consumer Credit Protection Act (protections against termination due to credit problems)[14]
13. Federal Water Pollution Control Act (protection for employees reporting violations)[15]
14. Judiciary and Judicial Procedure Act (protections regarding jury duty and related assistance to the courts)[16]
15. and specific industry protections such as the Railway Labor Act[17] and Railroad Safety Act.[18]

In addition to the federal statutes, numerous states have made inroads into the employment at-will doctrine through the enactment of state statutes affording protection to employees. As an

---

[7] 29 U.S.C.A. § 794.
[8] Discussed infra.
[9] 29 U.S.C.A. § 1140–1141.
[10] 38 U.S.C.A. § 2021(b)(1), 2024(c).
[11] 29 U.S.C. § 651–678.
[12] 42 U.S.C.A. § 7622.
[13] 5 U.S.C.A. § 7513(a).
[14] 15 U.S.C.A. § 1674(c).
[15] 33 U.S.C.A. § 1367.
[16] 28 U.S.C.A. § 1875.
[17] 45 U.S.C.A. § 152(3).
[18] 45 U.S.C.A. § 441(a) and (b)(1).

example, the state of California has enacted the following statutes providing protection to employees:

1. anti-discrimination laws (CGC § 3502)
2. protection of political freedom (CGC § 1101, et. seq.)
3. jury duty protection (CGC § 230)
4. polygraph restrictions (CGC § 637)
5. protection of military personnel (CGC § 394)
6. whistle-blowing statute (CGC § 10545)
7. antireprisal statute (CGC § 6400, et. seq.)
   a. forbids termination because of garnishment (CGC § 2929)
   b. forbids termination for reporting workers compensation injury or illness (CGC § 132a)

In recent years, the state courts have additionally developed exceptions to the rigid employment at-will doctrine. Many state courts have found that when an individual is terminated from his/her employment in contravention of some substantial public policy principle, such as refusing to commit a crime, that the public benefit far outweighs the employers right to terminate the employee. Most state courts recognize exceptions from the employment at-will doctrine for the following reasons:

1. upholding the law
2. refusal to commit an unlawful act (such as manipulating air samples)
3. reporting violations of the law (whistle blowing)
4. exercising a statutory right (such as religious preference)
5. performing a public obligation (such as jury duty)

Some state courts have gone even further in finding exceptions to the employment at-will doctrine. In some states, a covenant of good faith and fair dealing between employer and employee has been found in which a violation thereof could sustain a private cause of action against the employer.[19] The early cases finding a covenant of good faith and fair dealing usually involved an inherent wrong, such as discrimination prior to the federal regulations[20] or terminating the employee for monetary gain.[21] Later cases suggested that an implied covenant could even be inferred from an employer's policies rather than a specific act against the employee.[22]

Under the "implied-in-fact" contract theory, several state courts have found that an employer's personnel manual or handbook could constitute a contractual obligation thus removing the employer-employee relationship from the employment at-will status. The most publicized case to date is the decision by the Michigan Supreme Court that held that a provision in the employer's personnel handbook that the employer voluntarily adopted and distributed to employees stating that discharge would be for good cause protected employees from termination in the absence of a showing of good cause by the employer.[23]

Two other theories that have been recognized by some state courts is the promissory estoppel exception and the tortious discharge exception. Under the promissory estoppel doctrine, an individual has a reasonable reliance on the promise of another and when the promise is broken, the promisor is liable to the promisee. In the employment setting, a Minnesota court found that an employer who offered a position to an individual on which the individual relied and resigned from his current position only to find out that the offered position was being abolished was liable under the doctrine of promissory estoppel.[24] Under the theory of tortious discharge, a few courts have found that the circumstances in which an individual was terminated were so outrageous that the termina-

---

[19]Monge v. Beebe Rubber Co., 114 N.H. 130, 316 A.2d 549 (1974)

[20]Id.

[21]Fortune v. National Cash Register Co., 363 Mass. 96, 364 N.E. 2d 1251 (1977).

[22]Cleary v. American Airlines, Inc., 111 Cal. App. 3d 443, 168 Cal. Rptr. 917 (1981).

[23]Toussaiant v. Blue Cross & Blue Shield of Michigan, 408 Mich. 579, 292 N.W. 2d 880 (1980).

[24]Grouse v. Group Health Plan, Inc., 306 N.W. 2d 114 (Minn. 1981).

tion itself gave rise to a cause of action under intentional infliction of emotional distress.[25]

Although the basic foundational rule for employment in the United States is the employment at-will doctrine, the new laws on the federal and state level in addition to the court exceptions have made significant inroads in eroding this doctrine. Given the current trends, it is anticipated that the impact of the doctrine of at-will employment will continue to deteriorate at a significant rate.

REFUSAL TO PERFORM UNSAFE ACTS

Of particular importance to fire fighters are the protections afforded in the area of workplace safety and health. The general rule is to perform the work assigned by the employer and grieve the issue at a later time. This general rule can be extremely dangerous when the health and safety of the individual is in question.

The main case in the area of refusal to perform hazardous work is *Whirlpool Corp. v. Marshall*.[26] In this decision by the U.S. Supreme Court, the court found that, "As a general matter, there is no right afforded by the (OSH) Act, which would entitle employees to walk off the job because of potential unsafe conditions at the workplace."[27] However, the Court found that the intent of the OSH Act "does not wait for an employee to die or become injured" before becoming effective. Additionally, the Court found that "nothing in the Act suggests that those few employees who have to face this dilemma must rely exclusively on the remedies expressly set forth in the (OSH) Act at the risk of their own safety. But nothing in the Act explicitly provides otherwise. Again the background of legislative silence, the Secretary (of Labor) has exercised his rulemaking power under 29 U.S.C. § 657(g)(2) and *has determined*

---

[25]See, M.B.M. Company, Inc. v. Counce, 596 S.W. 2d 681 (Ark. 1980); Bodewig v. K-Mart, Inc., 635 P.2d 657 (Or. App. 1981).

[26]445 U.S. 1 (1980).

[27]Id.

*that, when an employee in good faith finds himself in such a predic-
ament, he may refuse to expose himself to the dangerous condition,
without being subjected to subsequent discrimination by the
employer.*" [28] In essence, this decision has provided employees the
right to refuse to perform hazardous work without fear of termination or retaliation by the employer through the providing of the
protections afforded under the OSH Act. (Note: the complete case
is provided to the reader in the case study section of this chapter.)

As part of the OSH Act enforcement scheme, Congress created a
special right of action for employees against whom employers discriminate because the employee exercised his/her rights under the
OSH Act.[29] This provision states that no person shall discharge or
in any manner discriminate against an employee because such
employee has filed any complaint or instituted or caused to be
instituted any proceeding under or related to this Act or has testified or is about to testify in any proceeding or because of the exercise by such employee on behalf of himself or others of any right
afforded by this Act.[30]

In addition to the above protections currently afforded under the
OSH Act, several bills are currently pending before Congress to
modify or reform the Occupational Safety and Health Act of 1970.
Under the vast majority of the proposed bills, employees will be
provided additional rights and protections in the area of workplace
safety and health.[31]

DISCRIMINATION

Title VII of the Civil Rights Act of 1964[32] and the new Civil Rights
Act of 1991[33] forbid discrimination in all areas of the employer-

---

[28]Id. (emphasis added)

[29]29 C.F.R. § 11(c).

[30]29 U.S.C. § 660(c)(1).

[31] See, e.g., Comprehensive Occupational Safety and Health Reform Legislation, supra.

[32]42 U.S.C.A. §2000E–2000E-17.

[33]Supra.

employee relationship when based upon race, color, sex (including pregnancy), religion, or national origin. The general purpose of Title VII of the Civil Rights Act was to require the removal of artificial, arbitrary, and discriminatory barriers to employment.

Of importance to public sector fire fighters is the fact that Title VII of the Civil Rights Act of 1964 was amended in 1972 by the Equal Employment Opportunity Act[34] extending coverage of the Civil Rights Act to employees of state and local governments and governmental agencies. The 1972 amendments also included special provisions requiring the federal government to comply with the Civil Rights Act provisions to personnel actions.

The federal agency charged with administration and enforcement of the Civil Rights Act and amendments is the Equal Employment Opportunity Commission.[35] A fire service organization falls under the jurisdiction of Title VII of Civil Rights Act of 1964 if

1. the organization has 15 or more employees
2. the employees have been employed for each working day in each of twenty or more calendar weeks in the current or preceding calendar year
3. the organization is engaged in an industry affecting commerce[36]

Title VII provides that it is an unlawful employment practice for fire service organizations or and employer meeting the definition to:

1. fail or refuse to hire or to discharge any individual, or otherwise to discriminate against any individual with respect to his or her compensation, terms, conditions, or privileges of employment, because of such individual's race, color, religion, sex, or national origin

---

[34]PL 92-261 (Mar. 24, 1972) 86 Stat. 103.

[35]42 U.S.C.A. § 2000e-4-5.

[36]42 U.S.C.A. § 2000e(b).

2. limit, segregate, or classify his or her employees or applicants for employment in any way that would deprive or tend to deprive any individual of employment opportunities or otherwise adversely affect his or her status as an employee, because of such individual's race, color, religion, sex, or national origin.[37]

Title VII of the Civil Rights Act extends not only to employers but also to labor unions, employment agencies, and labor-management training committees. As with most federal regulatory schemes, Title VII also contains and anti-retaliation provision where protections are afforded for individuals who report unlawful employment practices or filing a charge of discrimination.[38]

Of particular importance to fire fighters is the method through which to file or defend a Title VII claim. As noted above, the Equal Employment Opportunity Commission (known as the EEOC) is the governing federal agency. In many jurisdictions, state human rights commissions or other state agencies may be charged with the same or similar responsibilities.

In the event of an alleged unlawful employment practice by a fire service organization against a fire fighter in a state or locality that does not have a state statute or other law, the fire fighter must file a charge with the local office of the EEOC *within 180 days* from the occurrence of the discriminatory act.[39]

The charge can be filed with any EEOC district or area office or at the EEOC office in Washington D.C. If the charge is filed beyond the 180 day limitation, the claim may be lost. If the violation is continuing in nature, the 180 day time limitation may be considered inapplicable.[40] In states or localities with its own anti-discrimination laws and enforcement agencies, the time limitations

---

[37] 42 U.S.C.A. § 2000e-2(a).

[38] 42 U.S.C.A. § 2000e-3(a).

[39] 42 U.S.C.A. § 2000e-5(e).

[40] *Belt v. Johnson Motor Lines, Inc.*, 458 F.2d 443 (CA-5, 1972).

may be extended up to 300 days in which the state agency may file the charge with the EEOC.[41]

An EEOC charge must be in writing and must identify the organization or individual against whom the allegations are directed, identify the alleged unlawful conduct, and must be signed and verified.[42] Upon the filing of the charge, a fact finding conference is usually scheduled by the EEOC. The fact-finding conference is a face-to-face meeting between the parties. It should be noted that the fact-finding meeting is "on the record" and attorneys may be present but are limited to an advisory role and are not permitted to speak on behalf of the client or to cross-examine the other party.

Usually within 120 days following the completion of the investigation by the EEOC, the EEOC issues a determination. The determination can take two forms, namely reasonable cause determined or no reasonable caused determined.[43] A "no cause "determination signifies that the EEOC concluded that the alleged discrimination did not occur. A "cause" determination indicates that the EEOC believes that the alleged discrimination did occur. In the case of a "no cause" determination, the EEOC is required to provide the charging party with a "right to sue" notice through which the party may pursue the action in district court. A Title VII action must be filed in district court no later than 90 days following the issuance of a right to sue notice.

A finding of "cause" by the EEOC, the charging party may elect to either pursue the matter through the EEOC or may request a "right to sue" notice from the EEOC. It should be noted that the charging party may also request a "right to sue" notice from the EEOC if the investigative process goes beyond 180 days from the filing date prior to a determination. Given the current backlog of charges before the EEOC, this method is often used to accelerate the process by charging parties.

---

[41]42 U.S.C.A. §2000-5(e).

[42]29 CFR §1601.9. (Note: Verification may be satisfied by the administration of an oath at the time of filing. EEOC Compliance Manual §2.4.).

[43]EEOC Compliance Manual, § 40.

If the charging party elects to pursue the matter through the EEOC's administrative process following a determination of "cause," the next step is a conciliation session. In this meeting, the parties will attempt to settle their differences with the help of the EEOC. If an agreement is reached, the parties will enter into a written conciliation agreement.[44] If conciliation is not successful, the EEOC will send written notification that conciliation is terminated and send a right to sue notice to the charging party.

Either the charging party or the EEOC or both may pursue an action in federal district court to vindicate Title VII rights. The EEOC, in its discretion, may bring an action or decline to bring an action and defer to the charging party. Parties bringing a Title VII action are not limited in their election of remedies. Injunctive relief, back pay, reinstatement to the job, and attorney fees are frequently awarded. Class actions may be initiated in Title VII suits. Prior to the Civil Rights Act of 1990, all Title VII trials were *de novo* but now jury trials are permitted.

The defenses that may be available to a fire service organization in defending a Title VII charge include business necessity, bona fide occupational qualifications (known as BFOQ), a qualified seniority system, job testing, statistical data, "for cause" defense, and basic jurisdictional defects defenses. Additionally, in certain circumstances, specific defenses such as the national security defense, reliance on EEOC opinion defense, and religious preference defense, among others, may be used. Once the charging party meets their initial burden, the burden of proof falls upon the charged party to disprove the allegations or provide an appropriate defense.

DEMOTION AND DISCIPLINE

Employee discipline, the grounds for and procedures used, are an important topic in the American workplace. First and foremost,

---

[44]29 CFR § 1601.24(a); EEOC Compliance Handbook, § 40.8.

there is no one consistent rule in which all businesses and organizations are required to develop and administer workplace rules. As long as the workplace rules comply with the applicable federal, state and local laws, an employer may develop any type of workplace rule necessary, or no rules at all, and enforce the rules in any manner consistent with the applicable laws. In many organizations, the management group has developed specific rules and regulations through which to ensure consistency and accountability in the workforce. In other organizations, the employees have been empowered to develop their own rules and regulations. In either circumstance, problems usually arise when an individual allegedly "broke the rule" and disciplinary action is the remedy.

In a progressive disciplinary system, a series of progressively severe disciplinary measures are prescribed for each level of offense. For example, the first offense may be a verbal warning, the second offense is a written warning, the third offense is a three day suspension without pay, and the fourth offense is termination. In most progressive disciplinary systems, disciplinary action up to and including termination may take place on the first offense depending on the severity of the act. Most disciplinary systems require some type of recordkeeping system to ensure accuracy.

Disputes regarding disciplinary actions normally result because the party receiving the disciplinary action either believes that he/she is not guilty of the alleged offense or the disciplinary action imposed by management does not correlate to the alleged guilty act. The method for resolving these discrepancies runs the range from no appeal through a formal grievance procedure. In most circumstances, an employee cannot appeal a disciplinary action in a court of law or with an administrative agency until such time as the employee has severed the employment relationship. The initial step in most disciplinary actions beyond the verbal warning stage requires some type of investigative interview. In this step, the human resources manager or other designated individual within the management team conducts an investigation of the incident. This step can be one as simple as checking an attendance sheet to one as complicated as what was said in a charge of insubordination. Under the *NLRB v. J. Weingarten, Inc.* decision,[45] if an employee reasonably believes that the investigative interview might result in disciplinary action, the employee is entitled to have a co-worker or other representative present during the interview. In a union shop, this right of representation is usually required by the collective bargaining agreement and the union stewart or other representative is present during the interview. Under the *Weingarten* decision, employees in a non-union setting also have the right to have a representative or other employee present during the interview.

Following the investigative interview, the employee in question may be permitted to return to the job pending a final determination or may be suspended, with or without pay, for the duration of the investigation. Upon completion of the investigation, the employee is notified of the determination and the subsequent disciplinary action. This notification can be in the form of a face-to-face notification or, in the case of a termination, is often done via mail. It should be noted that with the substantial rise of workplace violence situations, appropriate precautions should be taken to safeguard individuals and the workplace in cases of face-to-face involuntary terminations. Given the volatility of the circumstances where an individual is losing his/her employment, often the management team member who is the bearer of the bad news, individuals who may have provided information detrimental to the individual being terminated, and even the workplace itself may be susceptible to violent outbursts by the terminated employee.

If the individual is represented by a union organization, a grievance procedure is normally set forth in the collective bargaining agreement. Grievance procedures are varied and may involve anywhere from a one time "hearing" before the company representative to a formal hearings before one or more outside arbitrators and a substantial number of appeal steps. In some non-union settings,

---

[45]420 U.S. 251 (1975).

various forms of re-evaluation steps may be included in the company handbook or established by policy.

PUBLIC SECTOR DISCIPLINE

For fire fighters employed in the public sector, the general rules regarding disciplinary actions are different than the private sector because of the Civil Service regulations.Additionally, disciplinary procedures between departments may be significantly different. Fire departments who have joined a labor union may have additional or different procedures as a result of the contractual obligations negotiated in the collective bargaining agreement.

In the public sector, disciplinary action can only be taken for violations of law, charter provisions, civil service regulations, and fire department rules. Disciplinary actions are usually delineated in accordance with the offense and each offense is usually designed with specificity. The power to remove a fire fighter in the public sector normally lies with the officer, board of directors, or tribunal as specified by the applicable statute.[46] For volunteer fire fighters working within the public sector, the rules for disciplinary action are usually the same as those set forth for paid fire fighters. In some states, specific statutes have been promulgated that determine the acts that may result in disciplinary action and the level of such discipline. Municipalities, counties, and other forms of local governments also usually have ordinances that specify the levels of disciplinary action that can be taken in accordance with the offense. In these situations, the highest ranking city or county official who has been vested with the appropriate authority usually makes the final determination as to the level of disciplinary action.[47]

Some fire departments are in a unique position in that the department serves at the pleasure of the city council or other governing board. Courts have held that entire fire departments can be termi-

---

[46]See, e.g., Nelson v. Baker, 112 Or. 79, 277 P. 301 (1924).
[47]See, e.g., Miller v. Town of Batesburg, 257 S.E. 2d. 159 (S.C. 1979).

nated by simply repealing the ordinance or law that created the fire fighters' positions. However, the court has found that fire commissioners or others vested with the authority cannot circumvent the law by terminating the entire department and then immediately reappointing all but the few fire fighters that were not wanted in the positions. Additionally, the placement of the fire department under the civil service system has been held by at least one court to not create vacancies in the positions already filled at the time the civil service act was adopted.

The rights of fire fighters in the public sector in the area of discipline are substantially different than those in the private sector. First, although the governmental entity has the ability to prescribe specific rules of conduct, each prescribed rule may not violate an individual's constitutional rights.

For example, if a fire fighter exercises his Fifth Amendment right to remain silent, this exercise of his constitutional right cannot serve as the basis for discipline.[48] Additionally, an individual may not be required to waive his/her constitutional rights as a condition of employment. Second, any violation that may result in disciplinary action must be as a result of a failure to comply with a prescribed rule and regulation and not an error in judgment. For example, simply being arrested cannot be the only reason for disciplinary action.[49] However, a conviction when the crime, such as a conviction for arson, can be in violation of a prescribed rule and thus grounds for involuntary termination.

One controversial area is discipline for "conduct unbecoming a member of the fire service." Discipline under this category is subjective at best and has been held by several courts to be unconstitutional unless the fire fighter was provided sufficient notice as to what conduct was required or what conduct was prohibited. In *Danner v. Bristol Township Civil Service Board*,[50] the Pennsylvania Supreme Court defines unbecoming conduct as "any conduct

---

[48]Garvin v. Chambers et. al., 195 Cal. 212, 232 p.696 (1924).

[49]See, e.g., Danner v. Bristol Twp. Civil Service Comm., 440 A.2d 701 (Pa. Cmnwlth 1982).

[50]Id.

that adversely affects the morale or efficiency of the bureau to which he is assigned. It is indispensable to good government that a certain amount of discipline be maintained in the public service. Unbecoming conduct is also conduct that has a tendency to destroy public respect for municipal employees and confidence in the operation of municipal services. It is not necessary that the alleged conduct be criminal in character nor that it be proven beyond a reasonable doubt . . . It is sufficient that the complained of conduct and its attended circumstances be such as to offend publicly accepted standards of decency."

Disciplinary actions under the category of "conduct unbecoming a member of the fire service" has been upheld in such circumstances as possession of controlled substances, possession of stolen property, and even witnessing sale of controlled substance by his wife in their home.

The grievance or appeal process in the public sector is normally established by the civil service regulations or collective bargaining agreement. In most circumstances, the fire fighter who received the disciplinary action is entitled to a hearing on the matter and often a subsequent hearing before an impartial arbitrator. Final disposition of such disciplinary matters normal lies in the courts in the public sector.

GRIEVANCE HEARINGS AND APPEALS

Grievance hearings or appeal hearings following the issuance of disciplinary action (see Fig. 10.1) run the spectrum from very informal hearings to hearings before a court of law. Fire fighters are strongly encouraged to acquire a working knowledge of the established grievance or appeal procedures within your department. In a majority of the grievance hearings, the burden of proof is on the party who has received the disciplinary action. Fire fighters should be aware that legal representation is usually permitted and a record of the proceedings is acquired for future use. The call-

EEOC CHARGE

(PLEASE PRINT OR TYPE)

APPROVED BY OMB	CHARGE OF DISCRIMINATION	CHARGE NUMBER(S) (AGENCY USE ONLY)
3046-0011 Expires 12/31/83	IMPORTANT: This form is affected by the Privacy Act of 1974; see Privacy Act Statement on reverse before completing it.	☐ EEOC

Equal Employment Opportunity Commission

NAME (Indicate Mr., Ms. or Mrs.)	HOME TELEPHONE NUMBER (Include area code)
STREET ADDRESS	
CITY, STATE, AND ZIP CODE	COUNTY

NAMED IS THE EMPLOYER, LABOR ORGANIZATION, EMPLOYMENT AGENCY, APPRENTICESHIP COMMITTEE, STATE OR LOCAL GOVERNMENT AGENCY WHO DISCRIMINATED AGAINST ME. (If more than one list below).

NAME	TELEPHONE NUMBER (Include area code)
STREET ADDRESS CITY, STATE, AND ZIP CODE	
NAME	TELEPHONE NUMBER (Include area code)
STREET ADDRESS CITY, STATE, AND ZIP CODE	

CAUSE OF DISCRIMINATION BASED ON MY (Check appropriate box(es))

☐ RACE   ☐ COLOR   ☐ SEX   ☐ RELIGION   ☐ NATIONAL ORIGIN   ☐ OTHER (Specify)

DATE MOST RECENT OR CONTINUING DISCRIMINATION TOOK PLACE (Month, day, and year)

THE PARTICULARS ARE:

(E5935)

FIGURE 10.1 Equal Employment Opportunity Commission charge form

ing of witnesses is usually permitted and the standard rules of procedure are followed.

The following format is usually used in a grievance hearing or arbitration hearing:

1. Opening statements—each party is usually permitted to provide a brief synopsis of their side of the situation.
2. Direct examination of witnesses—The aggrieved party is usually permitted to call witnesses to substantiate his/her side of the situation. It should be noted that arbitrators often require that a list of the witnesses to be called be submitted to the arbitrator and opponent in advance of the hearing.

3. Cross examination of the witnesses—each party is normally provided the ability to cross examine the opponent's witnesses. Leading questions are normally permitted in cross examination.

4. Presentation of "hard" evidence—each party is usually permitted to submit documents, photographs, and other evidence to the arbitrator for evaluation. Some arbitrators follow the rules of evidence adopted by the court but the vast majority of arbitrators permit documents and other "hard" evidence to be freely presented.

5. Closing statement—each party is usually permitted to present a final statement summarizing the case or emphasizing certain points of the case.

Preparation is essential for any disciplinary hearing or appeal from a disciplinary action. The aggrieved fire fighter must clearly show the arbitrator or judge the errors in the reasoning for the disciplinary action. The information presented to the arbitrator or judge should be clear and concise and based on objective evidence, if possible. Opinion, subjective analysis, and hearsay evidence should be maintained at a minimum. In a disciplinary proceeding, unlike a court of law, the opponent has the ability simply to remove the disciplinary action with little or no penalty at any time prior to the hearing. The basic reason why most disciplinary actions go to the grievance stage is because the management group believes that the disciplinary action was justified. The aggrieved fire fighter must show, in a clear and convincing manner to the arbitrator or judge, that the disciplinary action is inappropriate and should be overturned.

DRUG AND ALCOHOL USAGE/SCREENING

There is little doubt that the drug and alcohol problems being experienced in our society are spilling over to affect members of the fire service. Fire departments have a keen interest in identifying individual fire fighters who have a drug or alcohol problem in order to safeguard the individual, the fire fighters working side-by-side, the fire department, and the community. The basic problem in the area of drug and alcohol screening is the balance between the individual fire fighter's privacy rights and the fire department's need to know in order to protect those who may be affected by an impaired fire fighter's actions or inactions.

Drug and alcohol testing, usually part of an overall medical screening program by most fire service organizations, has evolved into a major area of controversy. With the major changes in technology, one blood sample taken for the purposes of screening for drugs or alcohol in the hiring process or at a yearly examination can also be used to test for HIV or even directed to the basic DNA level to identify genetic deficiencies. Does a fire department have a right to know what is in the blood system or the molecular make-up of their fire fighters? Does a fire fighter have a privacy right to his/her bodily fluids? In the area of drug and alcohol testing, the laws are evolving in order to provide guidance to fire fighters and the fire service but the other areas of testing continue to unfold. In the recent case of *Anonymous Fireman v. City of Willoughby*,[51] the court upheld mandatory HIV testing of fire fighters and paramedics as part of their annual physical examination. The court rejected all constitutional challenges under the fourth, ninth, and fourteenth amendments regarding individual privacy. Conversely, five states have enacted laws prohibiting genetic discrimination identified through testing in employment.[52]

In the area of drug and alcohol testing, a distinct delineation must be drawn between the permitted testing in the private sector and the public sector. In essence, random testing without probable cause is permitted in public sector positions where safety is at issue while random testing is prohibited in the private sector.

---

[51]779 F. Supp. 402 (N.D. Ohio 1991)
[52]Oregon (1989); New York (1990), Wisconsin (1992); Iowa (1992); and Rhode Island (1992). See also, Mark A. Rothstein, Genetic Discrimination in Employment and the Americans With Disabilities Act, 29 Hous. L. Rev. 23 (1992).

Drug and alcohol testing usually takes place in the following circumstances:

1. Pre-hiring testing for drugs and alcohol
2. In the private sector, testing when probable cause, such as following a work related accident
3. In the public sector, safety sensitive positions, such as a fire vehicle driver
4. Required periodic medical examinations. (As noted previously, drug and alcohol testing is not precluded in any way by the Americans With Disabilities Act.)

Fire fighters should be aware that the type of testing performed and the chain of custody on the sample have a great bearing on the accuracy of the results. With regards to the type of testing, there are various types of tests that can be performed ranging from the simple color change tests to spectroanalysis. As a rule of thumb, the more expensive the drug testing, the more accurate the results. With the basic color change testing used in most hiring situations, the accuracy level is normally between 80 and 90 percent. With laboratory analysis, the accuracy level is significantly higher ranging from 95 to 99 percent. Another variable to consider with drug testing is the specific drug in which the test is designed to identify, i.e., if the test is designed to identify THC, the test may not detect qualities of cocaine in the sample. Testing for alcohol also has differences in accuracy levels depending upon the test. The most accurate type of testing for alcohol is an analysis of a blood sample. Chain of custody is an important consideration in any alcohol and drug testing procedure. Chain of custody is, in essence, insuring that the sample tested belongs to the appropriate person. In most testing programs, the sample is followed and recorded at every step of the procedure from the taking of the sample through final analysis. Failure to adequately maintain an appropriate chain of custody can result in monumental errors in the disciplinary system.

Prior to establishing a drug and alcohol screening program, most fire service organizations develop an extensive written program outlining every aspect of the testing program including disciplinary action. Most programs have established procedures in which fire fighters abusing drugs or alcohol can voluntarily identify themselves and seek help through a qualified rehabilitation program. Rehabilitation programs cover the spectrum from organization paid in-patient care through providing unpaid time off to seek assistance. When a fire fighter is identified through random or for cause testing, rehabilitation is usually offered as the initial course of action. Subsequent "positive" tests can result in further rehabilitation and/or disciplinary action up to and including termination.

Summary

Termination from employment is always a costly matter. Loss of employment by a fire fighter dramatically affects the fire fighter on various levels including the economic level, psychological level, social level, and in various other ways. Termination from employment also affects the fire fighter's family relationship. For the fire service organization, the costs include hiring costs, training costs, loss of experience and expertise, department morale, and numerous other losses. Termination of employment is the last in a series of steps in an attempt to motivate the individual to conform to rules and regulations specifically designed to ensure efficiency, safety, and other vital components of a fire service organization. Discipline should always be evaluated from an objective point of view and should only be administered following a fair and impartial evaluation. Termination of employment is, in essence, the workplace version of the death penalty and should only be used when all else has failed.

SELECTED CASES

(This case has been edited for the purposes of this text.)

GRIGGS v. DUKE POWER CO.
401 U.S. 424 (1970)
MR. CHIEF JUSTICE BURGER delivered the opinion of the Court.

This case is before the court to resolve the question of whether an employer is prohibited by the Civil Rights Act of 1964,Title VII, from requiring a high school education or passing of a standardized general intelligence test as a condition of employment in or transfer to jobs when (a) neither standard is shown to be significantly related to successful job performance, (b) both requirements operate to disqualify Negroes at a substantially higher rate than white applicants, and (c) the jobs in question formerly had been filled only by white employees as part of a long-standing practice of giving preference to whites.

This class action was brought by a group of incumbent Negro employees against Duke Power Company. All the petitioners are employed at the Company's Dan River Steam Station in Draper, North Carolina. At the time this action was instituted, the Company had 95 employees at this location, 14 of whom were Negroes; 13 of these are petitioners here.

The district court found that prior to July 2, 1965, the effective date of the Civil Rights Act of 1964, the Company openly discriminated on the basis of race in the hiring and assigning of employees at its Dan River plant. Of the five operating departments of the plant, Negroes were only employed in the Labor Department where the highest paying jobs paid less than the lowest paying jobs in the other four "operating" departments in which only whites were employed.

In 1955 the Company instituted a policy of requiring a high school education for initial assignment into any department except Labor. When the Company abandoned its policy of restricting Negroes to the Labor Department in 1965, completion of high school also was made a prerequisite to transfer from Labor to any other department. The Company added a further requirement for new employees on July 2, 1965, the date on which Title VII became effective. To qualify for placement in any but the Labor Department, it became necessary to register satisfactory scores on two professionally prepared aptitude tests, as well as to have a high school education. In September 1965, the Company began to permit incumbent employees who lacked a high school education to qualify for transfer from Labor or Coal Handling to an "inside" job by passing two tests. Neither test was directed or intended to measure the ability to learn to perform a particular job or category of jobs. The requisite scores used for both initial hiring and transfer approximated the national median for high school graduates.

The Court of Appeals concluded that a subjective test of the employer's intent should govern, particularly in a close case, and that in this case there was no showing of a discriminatory purpose in the adoption of the diploma and test requirements. On this basis, the Court of Appeals concluded that there was no violation of the Act. This appeal resulted.

The objective of Congress in the enactment of Title VII is plain from the language of the statute. It was to achieve equality of employment opportunities and remove barriers that have operated in the past to favor an identifiable group of white employees over other employees. Under the Act, practices, procedures, or tests neutral on their face, and even neutral in the terms of intent, cannot be maintained if they operate to "freeze" the status quo of prior discriminatory employment practices.

The Court of Appeals' opinion and the partial dissent, agreed that, on the record in the present case, "whites register far better in the Company's alternative requirements" than Negroes. This consequence would appear to be directly traceable to race. Because they are Negroes, petitioners have long received inferior education in segregated schools and this Court expressly recognizes these differences. Congress did not intend by Title VII, however to guarantee a job to every person regardless of qualifications. In short, the Act does not command that any person be hired simply because he was formerly the subject of discrimination, or because he is a member of a minority group. What is required by Congress is the removal of artificial, arbitrary, and unnecessary barriers to employment when the barriers operate invidiously to discriminate on the basis of racial or other impermissible classification. The Act prohibits not only overt discrimination, but also practices that are fair in form, but discriminatory in operation. If an employment practice that operates to exclude Negroes cannot be shown to be related to job performance, the practice is prohibited.

Neither the high school completion requirement nor the intelligence tests required by Duke Power is shown to bear a demonstrable relationship to successful performance of the jobs for which it was used. The Court of Appeals held that the Company had adopted these requirements without any "intention to discriminate against Negro employees." However, good intent or absence of discriminatory intent does not redeem employment procedures or testing mechanisms that operate as "built-in headwinds" for minority groups and are unrelated to measuring job capability.

Nothing in the Act precludes the use of testing or measuring procedures; obviously they are useful. What Congress has forbidden is these devices and mechanisms controlling force unless they are demonstrably a reasonable measure of job performance. Congress has not commanded that the less qualified be preferred over the better qualified simply because of minority origins. What Congress has commanded is that any tests used must measure the person for the job and not the person in the abstract.

The judgment of the Court of Appeals is, as to that portion of the judgment appealed from, reversed.

Index